A NEW WORLD-SYSTEM

A New World-System: From Chaos to Sustainability examines the present crisis in the social and ecological environment that is producing profound, potentially catastrophic challenges to the planet and humanity and outlines a process for moving forward to address these critical issues.

This book is a cautionary interpretation of the present and vision for the future. Unlike other books on this or allied subjects that are focused singularly, Part 1 surveys the five major threats facing humanity today: climate change, inequality and poverty, new technologies, migration, and globalization. It approaches the challenge of integrating these phenomena into a global picture from a systems perspective rather than taking a purely reductionist approach to understanding what is occurring in the world today. Part 2 moves from identifying the problems to solving them, with chapters examining the ability of the present world-system to address these issues and outlining a process for action. The book concludes by discussing what could follow capitalism as a social organizing strategy and, perhaps more importantly, the consequences to the planet if we do not construct a new world-system.

This book is essential reading for students and scholars of sustainable development, climate change, environmental studies, rural and urban planning, environmental psychology, political economy, sociology, social policy, leisure studies, and environmental politics. More broadly, it is a vital resource for all those interested in building a sustainable society.

Donald G. Reid is University Professor Emeritus in the School of Environmental Design and Rural Development (SEDRD) at the University of Guelph in Canada. He is an international scholar whose work focuses on sustainable development, social planning, poverty, community development, leisure, and tourism. His earlier book, *Social Policy and Planning in the 21st Century: In Search of the Next Great Social Transformation*, was published by Routledge in 2017.

Routledge Studies in Sustainable Development

This series uniquely brings together original and cutting-edge research on sustainable development. The books in this series tackle difficult and important issues in sustainable development, including values and ethics, sustainability in higher education, climate compatible development, resilience, capitalism and de-growth, sustainable urban development, gender and participation, and well-being.

Drawing on a wide range of disciplines, the series promotes interdisciplinary research for an international readership. The series was recommended in the *Guardian*'s suggested reads on development and the environment.

Poverty and Climate Change
Restoring a Global Biogeochemical Equilibrium
Fitzroy B. Beckford

Achieving the Sustainable Development Goals
Global Governance Challenges
Edited by Simon Dalby, Susan Horton and Rianne Mahon, with Diana Thomaz

The Age of Sustainability
Just Transitions in a Complex World
Mark Swilling

A New World-System
From Chaos to Sustainability
Donald G. Reid

For more information about this series, please visit: www.routledge.com

A NEW WORLD-SYSTEM

From Chaos to Sustainability

Donald G. Reid

First published 2021
by Routledge
2 Park Square, Milton Park, Abingdon, Oxon OX14 4RN

and by Routledge
52 Vanderbilt Avenue, New York, NY 10017

Routledge is an imprint of the Taylor & Francis Group, an informa business

© 2021 Donald G. Reid

The right of Donald G. Reid to be identified as author of this work has been asserted by him in accordance with sections 77 and 78 of the Copyright, Designs and Patents Act 1988.

All rights reserved. No part of this book may be reprinted or reproduced or utilised in any form or by any electronic, mechanical, or other means, now known or hereafter invented, including photocopying and recording, or in any information storage or retrieval system, without permission in writing from the publishers.

Trademark notice: Product or corporate names may be trademarks or registered trademarks, and are used only for identification and explanation without intent to infringe.

British Library Cataloguing-in-Publication Data
A catalogue record for this book is available from the British Library

Library of Congress Cataloging-in-Publication Data
Names: Reid, Donald G., author.
Title: A new world-system : from chaos to sustainability / Donald G. Reid.
Description: Abingdon, Oxon ; New York, NY : Routledge, [2021] | Series: Routledge studies in sustainable development | Includes bibliographical references and index. |
Identifiers: LCCN 2020022877 (print) | LCCN 2020022878 (ebook) | ISBN 9780367611309 (hbk) | ISBN 9780367609672 (pbk) | ISBN 9781003104261 (ebk)
Subjects: LCSH: Sustainability. | Sustainable development. | Climatic changes. | Social change.
Classification: LCC HC79.E5 R4473 2021 (print) | LCC HC79.E5 (ebook) | DDC 304.2—dc23
LC record available at https://lccn.loc.gov/2020022877
LC ebook record available at https://lccn.loc.gov/2020022878

ISBN: 978-0-367-61130-9 (hbk)
ISBN: 978-0-367-60967-2 (pbk)
ISBN: 978-1-003-10426-1 (ebk)

Typeset in Bembo
by Apex CoVantage, LLC

This book is dedicated to Patricia, Sherry, Bill, Heidi, Ron, Nicole and Ethan.

CONTENTS

List of illustrations ix
Acknowledgements x
Preface xi

PART 1
Forces driving change 1

1 Introduction: humanity's next test 3

2 Climate change: the elephant in the room 15

3 Technology: the Trojan Horse 35

4 Poverty: the great unequalizer 56

5 Globalization: the eye of the storm 73

6 Migration: the clash of civilizations 91

PART 2
Processes of change 107

7 Recent change experiments 109

8 The context for social change 131

9	Theories of change	142
10	Planning for change	167
11	Meeting the challenge: from chaos to sustainability	192
	Epilogue	215

Index *219*

ILLUSTRATIONS

Figures

9.1	The world-system and component parts requiring alteration	147
9.2	Theories of biological and social adaptation	152
9.3	Theories and critical components of the path to the development of a new world-system	157
9.4	Forces of the life- and system-worlds	159
10.1	World-system change framework	178
11.1	Capitalism to predatory corporate capitalism	195
11.2	The present balance of the life- with the system-world	203

Tables

2.1	Fundamental conclusions on global warming	25

ACKNOWLEDGEMENTS

There are many people who helped guide this project. I first wish to thank my partner in life Patricia MacPherson for her continued encouragement and for her comments on the manuscript as it progressed through its many stages. I would also like to thank our Seville friends who provided hours of discussion leading to many of the ideas that eventually wound up in this book. To Karen, Rich, Cristina, Jimmy, Isabel, and Julio, I thank you for your friendship and inspiration. Finally, I want to extend my thanks to Reyes Lopez-Huerta, who shared her home in Seville with us where much of this book has been written.

PREFACE

The aim of this book is to identify and examine the present chaos in the social and environmental systems that is producing profound, potentially catastrophic challenges to the planet and to society. In order to address the crisis, this book proposes large-scale changes to the world-system that dominates our everyday life.

In an earlier book titled *Social Policy and Planning for the 21st Century: In Search of the Next Great Social Transformation* (Reid, 2017), I noted what I believed to be the major problems facing humanity and the issues that will determine the survival of the planet and human species in the future. Essentially, those boiled down to climate change and global warming, immigration, the development and control of new technology, poverty and unequal distribution of wealth resulting in extreme economic separation between the rich and the poor, and the issues of globalization in general. But, in one sense, these problems are a symptom of something much larger. A major inspiration leading to the writing of this book is the understanding that none of the problems cited here can be resolved independently but require a comprehensive and system-wide approach to their resolution. They are intricately linked, and solving one, independent of the others, may lead to making the rest worse if they are not tackled in unison and by not recognizing each one's role in the overall system. These issues reject a reductionist analysis and require a systems approach to their resolution. Resolving these portentous concerns must be accomplished within an overarching framework focused on transforming the present world-system in order to establish a more economically just and environmentally centred social organizing structure. The present world-system – predatory corporate capitalism – is not ideologically equipped to underpin the magnitude and kind of change that is required to rebuild the social architecture that leads to a stable physical and social environment.

This book is a cautionary interpretation of the present and vision for the future. On the surface it appears that the state of the world has never been better for most of us, even though it doesn't always feel like it. We are healthier, better educated, and less violent than ever before. Some of us are even wealthier than the generations that have come before us. There are still some parts of the developing world that don't enjoy that wellbeing however. And even in the affluent world, there remain a significant number of people suffering from poverty. There is still racism and division among the population as well. Some argue we are engaged in a culture war. Sometimes we are hit with events such as the Coronavirus (COVID-19) that reminds us that life is a tenuous journey.

Despite our general good fortune, there are some worrying clouds on the horizon. In fact, those clouds may already be swirling around us, and we are not paying enough attention to them. Humanity is encountering some very severe problems that are not being adequately addressed. Present-day society seems to be solely focused on the enhancement of wealth despite what those efforts are producing in the form of climate change, global warming, and social disruption. We have become a very me-oriented society, and that myopic view is leading us into social and environmental chaos.

Human social development is not a continuing, progressive, upward trajectory; there are ups and downs. There have been Dark Ages in addition to Renaissance. The Industrial Revolution, particularly the period after World War II, was a great leap forward, as it increased the standard of living for a huge number of people around the globe, particularly in the democratic West. A large middle class was generated in many parts of the world. The rise in affluence was felt even more momentously given that it occurred directly on the heels of an economic depression and world war. The recognition of the rise in affluence in the West does not ignore the many people in the world who have not been included in that economic good fortune. In fact, a goodly portion of the West's affluence has been built on the backs of the developing world. There are still far too many people around the globe who struggle each day to achieve enough calories to stay alive. Even in the affluent countries, the inequality that exists is becoming increasingly pronounced each day. In considering the larger picture, however, humans have fared well over their short existence on this planet. The advance of human progression is undeniable. Since the time of hunting and gathering to the advanced technological society we enjoy today, many substantial developments have occurred throughout the human story. Today's society has become complacent, however, and we take for granted the recent upward progression of society as a naturally occurring phenomenon, one that will continue long into the future. But our future is anything but secure – in fact, for many of us it appears to be stagnant or even in decline. As a reaction to this apparent breakdown in social cohesion, some of us seem to want to return to a time past, but it is doubtful that moving in a backward direction is even possible let alone desirable, or that it would produce more stability to everyday life than the present condition. In fact, most observers suggest that improving life and society is to be found in moving forward, not languishing in nostalgia for some long-ago period. Society's culture is dynamic and ever changing even though many societies greatly resist that change. This book takes as its starting point the position that human society has become too sanguine and is ignoring some very distressing problems that could produce dire consequences to us and the planet if we do not take strong actions to address them immediately. Modern society is at a critical turning point. It would not be hyperbole to suggest that the existence of humanity, perhaps all life on the planet, is under threat because of these issues. Despite the positive human development story that has played itself out since the arrival of hominids on the savanna, there are some dark clouds on the horizon that jeopardize that progressive story. In fact, we may be about to confront our very existence as a species.

Although the critical subjects identified in Part 1 of this book appear on the surface to lend themselves to individual solutions, they need to be linked together as a system of concerns and not as individually resolvable problems. Each is at a critical stage because they are part of a world-system that is now dysfunctional and requires remediation at the macro scale. The problems that humans face today are unlike any they have addressed before. They are of the variety that requires a new worldwide operating system if they are to be addressed satisfactorily. A new social architecture is much needed as we tackle the problems of the 21st century and beyond. Further, many of the problems that confront us are a result of the advances that were made by previous generations that brought us to the state of affluence we enjoy today. They are an outcome of the framework that ushered in and maintained the Industrial Revolution. So what humanity has constructed as an economic system has been positive in the past and produced a very affluent world and a productive and enjoyable lifestyle for most of us, but it has also produced unintended consequences that now threaten to destroy civilization and the planet on which we live.

In this book I have chronicled what I consider to be the most critical problems confronting humanity and discuss what needs to be done to address them. Certainly, there is no silver bullet on the horizon that can be employed to magically solve all the problems faced by our present society. Only a major transition of the social architecture at the world-system scale will be adequate to the task of getting humanity back on the road to environmental sustainability and social justice for all.

Part 1 of this book consists of original thought and an analysis of the work of experts in each of the subjects explored. To examine these subjects fully, many scientific and professional reports are reviewed, analyzed, and interpreted. Part 2 discusses what needs to be done and how to go about doing it in order to ameliorate the problems discussed in Part 1.

PART 1
Forces driving change

1
INTRODUCTION
Humanity's next test

The capitalist world-system is declining in effectiveness, and some might even say it is antithetical to present and future human development. Some (Wallerstein, Collins, Mann, Derluguian, & Calhoun, 2013; Streeck, 2016) argue that capitalism is not only in decline but will come to an end in the not-too-distant future. Few see it ending abruptly but becoming increasingly dysfunctional in maintaining the stability of the planet and social system over the near and long-term. It is likely to stumble along as the world-system lapses into greater dysfunction or until humanity collectively acts and adopts something more functional.

The idealized notion of capitalism has morphed into predatory corporate capitalism and may be entering a period of self-consumption. It flourished in the past because it rationed scarce resources in an organized manner, maybe what the world needed at the time, but that task is now complete, and humanity is confronted with a completely different set of problems. Corporate capitalism performed its original function without regard for the externalities it produced that severely damaged the environment, and without concern for establishing an equitable system for sharing wealth among the population. The capitalist system survives because it externalizes its waste onto the land, into the water and atmosphere, and pillages the world's resources at an unsustainable rate. Additionally, a few individuals now capture most of the fruits of humanity's labour. The problem of maximizing resource use has been solved. The problem our economies face today is not the lack in production of enough goods and services to satisfy human needs but with their overuse and the unequal distribution of the income produced by their production and consumption in the capitalist economy. The task for society is to create a suitable replacement for this destructive behaviour and achieving this objective will be no easy mission.

System change is being demanded by much of society. Frustrations on the part of many have bubbled to the surface and can no longer be ignored by the power structure. Millennials who will inherit this damaged planet are extremely concerned about environmental conditions and the politics that ignore the mounting social and environmental destruction. Politicians and those with power are attempting to slant the demand for change to their benefit without focusing on what is in the best interest of society. Conservative voices focusing on the past are calling for isolationist and nationalist policies as they disregard the destruction of the planet and the dysfunction of the social system. By contrast, some are calling for a new world-system that opens individual countries to a cooperative and comprehensive globalization that focuses on policies beyond purely economics. As Wallerstein states, '(t)he social reality in which we live and which determines what our options are has not been the multiple nation states of which we are citizens but something larger, what we call a world-system' (Wallerstein, 2004, p. 9). Although the trend today is to become isolationist, the issues that confront humanity are of a global, not local, nature. Their remediation will only be accomplished through a collective global effort. The recent crisis caused by the COVID-19 virus is testament to that fact.

Humanity has constructed a path that binds one nation to the rest through powerful economic forces, in addition to extensive travel patterns, rapidly developing technology, and a worldwide communications network. The world order in which we live today is a system with dependent and interacting parts that respond to one another. The focus of a new world-system must emphasize the relationship between and among those parts as much as on the parts themselves. However, it is the parts that are receiving attention today, and the relationship among them is of little or no concern to the present power structure. In fact, some politicians are blaming forces beyond their control for the inability of the middle and lower economic classes to maintain their standard of living and for the catastrophes confronting the environment. They argue that by closing their national system to the rest of the world, they can control or eliminate these problems. Blaming outside forces for internal problems is misplaced and will simply lead to further degradation in the system. Many of the problems, particularly environmental and social justice concerns, are worldwide issues that need to be addressed collectively by all nations. Any measure a nation takes to address these issues will have consequences for all other nations. So, a new world-system must focus not only on the content (i.e., social and environmental outputs) but also on relational processes (interaction among the parts of, and players in, the system).

The world suffers from a runaway economic system that is now in charge of society rather than the other way around. The financialization of the economy has created overly large financial companies to the point where some of the big corporations have been deemed too big to fail. This notion itself is contrary to capitalism's basic code of creative destruction (Schumpeter, 2016), so our present economic system seems no longer able to adhere to its own fundamental principles. The unprecedented size of both public and private debt is another indication of the emergency. Further, there has been a transfer of wealth from the bottom of society to the top through excessive tax cuts benefiting the wealthy, and the deinstitutionalization of the public enterprise by turning many long-standing functions of the government over to the private sector. Several programs originally thought to be the responsibility of government have been commodified and become part of the market economy, leaving many of those at the bottom of the social hierarchy without access to adequate service. Moving capital from the bottom of the social classes to the top and increasing the use of debt to make it appear that the economy is growing is a clear indication of the decay. To achieve the capital flows from the bottom to the top of society, the oligarchs have seized control of government and public policy through the application of their wealth to politics and the takeover of other vital public institutions (Mayer, 2016). The United States and many other western governments are no longer true democracies but plutocracies. Some are even showing signs of a surging authoritarianism.

It is clear to many that predatory corporate capitalism has run its course and needs alteration if not complete transformation, if civilization is to regenerate and thrive. That is not to say that we will need to abandon free markets entirely but place much less reliance on the economic marketplace for organizing humanity's social relations in the future than we do now. John McMurtry's 1999 book titled *The Cancer Stage of Capitalism* leads to the logical conclusion that capitalism is no longer healthy in light of the present-day needs of society and the planet. In fact, today's version of capitalism may be perpetuating the present crisis.

The world seems to be in greater political and environmental turmoil today than since the launch of the modern era. That said, I suspect that all societies in the past believed they too were transecting difficult times, although today's confrontations seem to be unusually daunting. Certainly, such recent historical events as the Great Depression of the 1930s and World War II caused considerable consternation and disorder among the population, but, for the most part, those societies demonstrated social solidarity and collective determination to overcome the disorder that threatened them. Today's society seems more divided than ever before. Although times were tough, western society didn't face the threat of structural collapse as it does today. The problems facing human society and the planet today must be considered a distinctive demarcation point in human history, one of those rare occasions when humanity is confronted with its very survival. Not only are there physical and social threats, such as global warming, exaggerated inequality, and other social ills associated with economic separation, but also challenges that threaten the foundations of the social system and basic way of life contained within democracy.

Many western liberal democracies risk the threat of sliding into authoritarianism or even fascism. This has come about by a 'deep sense of alienation, loneliness, and anxiety that haunts the United States at the present moment', and 'the path we are on will lead to more misery and conflict' (Giroux, 2018, pp. 32, 48). Although Giroux was speaking chiefly about the United States, his comments are applicable to other liberal democracies around the world as well.

Liberal democracies in the 21st century have abandoned all sense of cooperation and collectivism in human relations and have degenerated into individualistic narcissism. This shift in paradigm has produced profound anomie and mass anxiety within society. Eisenstein reasons that:

> In the face of an ecological, financial, social, and health crisis that isn't going away, our tools – political, technological, and cognitive – are revealing themselves as impotent. As that happens, the belief systems that embed those tools lose the gloss we call 'reality'. Our defining narratives are coming apart at the seams. Our society is based upon competition and anxiety in part because these are implicit in our basic understanding of the universe. To forge a new psychology – and, collectively, a new society – that is not underpinned by anxiety will therefore require a new conception of self and life, and therefore of science and the universe. . . . this mode of being [competitive narcissism] will come to an end, to be replaced by a profoundly different sense of the self, and a profoundly different relationship between human and nature.
>
> *(Eisenstein, 2007, pp. 10, 19, 85)*

Many social analysts, commentators, and pundits are concerned that the corporate and predatory capitalism in which our world is now fully entrenched is leading western democracies toward authoritarianism. There is no doubt that the United States and many other western nations such as Turkey and Hungary are being governed by populist and nationalist regimes and perhaps headed toward authoritarianism, if they aren't there already.

As its point of departure, this book posits that the cooperative and collective society has given way to a predatory corporate capitalist culture that focuses on profit maximization without concern for the cost to the social fabric or the physical environment. The damaging effect this deterioration has and will continue to perpetuate on society is profound, perhaps even life threatening. If drastic changes to the global social system are not initiated soon, the damage will be catastrophic.

The wanton disregard for the welfare of society is being felt most profoundly by the lower socio-economic segment of society. The lower and middle economic classes are demonstrating their frustration at being left behind by today's economic and socially deteriorating trends and are becoming angrier about it each day. Evidence of this anger is demonstrated by the demonstrative protests stretching around the world by the Black Lives Matter movement. Instead of new ideas, today's politicians are offering worn-out programs and ideologies from the past that didn't work for everyone then and are not likely to work now. What's more, the marginalized in society are being offered emotional clichés, not remedies that will change their situation for the better. Timothy Snyder in his book *The Road to Unfreedom* is worth quoting at length when he speaks about the politics of inevitability:

> The politics of inevitability is the idea that there are no ideas. . . . The cliché of the politics of inevitability is that 'there are no alternatives'. To accept this is to deny individual responsibility for seeing history and making change. . . . The capitalist version of the politics of inevitability, the market as a substitute

for policy, generates economic inequality that undermines belief in progress. As social mobility halts, inevitability gives way to eternity, and democracy gives way to oligarchy. An oligarch spinning a tale of an innocent past, perhaps with the help of fascist ideas, offers fake protection to people with real pain. Faith that technology serves freedom opens the way to his (sic) spectacle. As distraction replaces concentration, the future dissolves in the frustrations of the present, and eternity becomes daily life. The oligarch crosses into real politics from a world of fiction, and governs by invoking myth and manufacturing crisis.

(Snyder, 2018, pp. 19–20)

Snyder presents a bleak picture of where we are headed. He believes we are already well down the road to political inevitability and cautions that this is a world without hope.

All signs point to the need for drastic change in our world-system, yet society doesn't seem to hear the call or, if it does, doesn't give the impression it can address the pending crisis with new ideas that have a chance of being successful. The environmental, economic, and social systems continue to spiral downward for a large segment of the population, and the solutions proposed by the political system don't appear to address those problems with any sense of urgency or effectiveness. The present political system and its proposed solutions are bound by the current world-system, and until that boundary is breached, the social and environmental conditions will continue to slide into more and more turbulence and anomie.

As a vital part of identifying and addressing the major issues confronting society today, this book identifies social theories that can be useful in establishing a framework for changing the world-system under which we now exist. Theories provide us with a lens for interpreting the world in which we live. They can help us make sense out of what appears to be chaos. Subsequent to identifying and outlining important social theory that helps us understand our world, this book will provide thoughts on the methodologies, processes, strategies, and actions that can be employed to establish a new world-system, one that will be compatible with environmental sustainability and more appropriate for continuing human and social life on this planet.

The present world-system needs to be replaced with a new strategy if the current and looming problems confronting the planet and society are to be addressed successfully. The basic aim of this book is to explore the idea of world-system change. It examines the present crisis in the social and ecological environment that is producing profound, potentially catastrophic, challenges to the planet and the social system. In addition to outlining the immediate problems our society is experiencing (Part 1) but sleepwalking through, I examine the ability of the present world-system to address these dire concerns (Part 2).

The world is in crisis because the present world-system, capitalism, no longer serves society and the majority of its people. As predicted by Marx, the world's wealth is being concentrated in fewer and fewer hands, and the continual push for increasing profits is causing severe environmental damage, including changing the climate and temperature of the planet. Some authors, such as Giroux, Harvey, and Snyder for example, view the extreme push for profit at all costs to be no longer healthy for democracy or the planet. We have drifted into a predatory corporate capitalism that no longer benefits all citizens but only the wealthy few among us. Historically, the human community has outlived and transitioned through many economic organizing strategies. It is now time to move on to a new economic formation and world-system, one that is more in harmony with the environment and the basic needs of humanity. I have no confidence that the present predatory, corporate, capitalist system has the capability or the will to make such change. That said, the first issue to be addressed is to answer the question of whether there is, in fact, such a thing as a world-system or whether that notion is simply myth and unnecessary to the analysis and formation of environmental and social policy for the future. One, if not the most respected proponents of the reality of a world-system and its power over life on the planet is Immanuel Wallerstein. Wallerstein states:

> We have been arguing that the social reality within which we live, and which determines what our options are has not been the multiple national states of which we are citizens but something larger, which we call a world-system. Proponents of world-systems analysis . . . have been talking about globalization since long before the word was invented – not, however, as something new but as something that has been basic to the modern world-system ever since it began in the sixteenth century.
>
> *(Wallerstein, 2004, p. 9)*

The basis for the present world-system is the constant need for increasing the accumulation of capital facilitated by new technology and the expansion of geographical, psychological, intellectual, and scientific frontiers that extended markets, allowing for constant economic growth at an accelerated rate. Before the present world-system began, others existed but perhaps on a much smaller scale. Hunting and gathering, agriculture, and the advent of industrialism powered by steam, denotes such previous systems. It is arguable that in previous times much of the world was undiscovered, so the world-systems of the past were not as truly worldwide or as complex as today's.

In an earlier work (Reid, 2017), I argued that human society is on the brink of, or immersed in, a great social transformation much like past societies underwent. For example, humans transformed their way of life from hunting and gathering to agriculture, from subsistence agriculture to feudalism, to mercantilism, to industrialization, and so on throughout time. Despite many obstacles and setbacks along the way, these transitions worked well for humanity, and society progressed positively throughout the ages. With the extreme change in global climate resulting in the earth's warming, and with the recent advancement and proliferation of digital technology, society is yet again faced with change equal to, or exceeding, the magnitude of the challenges overcome by previous societies. It is these tensions we feel throughout society today, and I'm not sure we are managing them well. Snyder (2018) certainly warns us about sleepwalking through history by not stepping out of our present box, not creating new ideas, and not selecting the positive choices that could form our future.

If a culture becomes static, it will no longer perpetuate human endurance, and the species will die out. Present society seems to be confused as to whether it wants to move forward socially or regress to an earlier time. At this point in human history, some of us seem to be longing for a return to a time past, but it is doubtful that moving in a backward direction is even possible let alone desirable, or that it would produce more stability to everyday life than the present condition. In fact, most observers suggest that improving life and society is to be found in moving forward, not languishing in nostalgia for some long-ago period.

What has been most notable about humanity throughout history is its ability to adapt to a continually changing social and ecological environment. Although social change has not been constant or even, humans have made great leaps forward as they adapted successfully to alterations in the environment at the most critical periods throughout history. Some of these major changes in human existence will be chronicled in detail later. This feature has been instrumental in their ability to flourish and become the dominant species on the planet. Change and adaptation is a constant process and it occurs despite great resistance put up by humans themselves. Humans don't appear to seek or like change despite our demonstrated ability to use it to our advantage. Without question, we have been a very successful species, until the present time at least.

The question of whether or not a completely new social organizing system needs to be considered as we enter the new age or would simply tinkering around the present structure making changes at the margins be sufficient? Given the domination of our culture by the present aberrant economic system and the damage to the climate and social justice system it has produced, tinkering at the margins would likely not be enough. We see big cracks in the system when we feel the anger and fear that grips many of our fellow citizens today or when we receive another jolting report on the climate by the Intergovernmental Panel on Climate Change (IPPC). The future does not seem all that bright for many people, perhaps for all of humanity, under the present world-system.

Recent changes to the physical environment in the form of global warming and social regression, including a distorted economic system producing a massively unequal wealth distribution system, have made a few of us very wealthy but many more of us poorer and at the expense of the ecosystem. The physical environment is deteriorating rapidly and needs rehabilitation. In addition to the deterioration of the environment and increasing economic dislocation, there is great angst in society today due to the large numbers of migrants worldwide and the clash of cultures migration is producing. These major factors are driving the need for another great leap forward in the social organizing system. Whether humanity can make the required social transition and overcome these problems is yet to be determined, although much hangs in the balance. The environmental and social problems with which this generation of humanity is confronted are daunting and a huge challenge, with humankind's very existence being put in question.

Collectively, society has within its power the ability to resolve present and potential catastrophic challenges through social transformation or, conversely, succumb to them and lapse into environmental and social chaos and disintegration. The magnitude of the challenge today only seems like a greater test than any of the historic human social transitions engaged previously because we are immersed in it and feel the anxieties the system has produced. While earlier transformations were mainly the result of new biological and mechanical technologies, producing an expansion in food production or population, this transformation is being pushed by the exponential increase of new digital technologies, unequal wealth distribution, and global warming. The original capitalist system, so important to supporting the early Industrial Revolution, has lost its utility as an organizing mechanism. The capitalism that fostered broad-based upward economic mobility that elevated the living conditions of most, if not all, of the population in the developed world has given way to predatory corporate capitalism that enriches only the few. In the United States, 1% of the population owns 35% of the country's wealth. In Canada, the 1% owns 37% of that country's wealth. This pattern is similar throughout the rest of the affluent world. According to Fukuyama, 'the 1% took home 9% of the GDP in 1970 but 23.5% in 2007' (Fukuyama, 2014, p. 617). These are dramatic numbers that show just how predatory and unbalanced the capitalist system has become. Middle and lower socio-economic class incomes in many developed countries have been stagnating, if not receding, over the last two decades, while the richest 1% of the population has been getting wealthier. The unequal sharing of resources has had a profound and debilitating effect on our democracy.

The consequence of this unequal distribution of the world's wealth has been a driving force in the movement toward authoritarian and totalitarian governments in many western democracies. Today, Turkey, Hungary, Poland, the Philippines, and Venezuela and some lesser states seem to be heading toward authoritarian regimes, if not there already. The present US administration acts like an authoritarian government by challenging the legitimacy of democratic institutions and engaging in nepotism, and other equally anti-democratic practices. Is this really the road humanity wishes to travel in the 21st century?

Because of the present nostalgia for how things used to be, the current social and economic system is faltering. Present-day society in North America, if not also in Europe and beyond, yearns to return to a time that is perceived to have been better for the majority than it is today. It is generally understood, however, that no society can retreat to days gone by but, instead, must look forward to new horizons and for new mechanisms for continuing social development. We require a new social and economic organizing system to match the requirements of our changing environmental and social reality, not one that venerates the past but doesn't address the critical problems of the present. Many of our fellow citizens sense they have been side-lined from the theatre of life and no longer feel there is a clear path forward or that they are in control of their future. No amount of tinkering around the margins of the present system will produce the results needed to alleviate the anxiety rampant in society today. The feeling of the population at large is, and perhaps rightfully so, that an elite group of individuals is controlling society and not doing so with the best interest of the public in mind but rather working only for their own benefit. Social scientists have termed this form of governing, plutocracy, rule by the rich. Plutocracy forms when the wealthiest of society usurp power, dominate government, and act in their own self-interest.

Plutocracy has been operative long before the present US administration came into power. The most recent version has been noticeable since the middle of the last century's globalization process. Globalization has been hijacked from the start by the wealthy and powerful who have shaped international trade and politics for the benefit of themselves and their corporations, not for uplifting the human condition in general. The basic guiding principle underlying the present version of capitalism is 'trickle down', which suggests that if those at the top do well economically, their good fortune will trickle down to those at the bottom. Not only is this theory now totally discredited but has caused those at the top to overreach and squeeze more out of the system than justified, even to the point of doing great harm to the entire social system and the planet. This was evident during the run up to the economic catastrophe of 2008 that resulted in the near collapse of the global economy. The supposedly self-adjusting capitalist system in the United States and other parts of the western world had to rely heavily on government to bail them out; otherwise, the entire economy would have collapsed, causing another great depression not unlike the 1930s. The economic system's inability to correct itself due to human-induced crisis has given us a glimpse of its frailty but we didn't seem to digest the lessons presented by that near collapse.

In this brave new world that is bound to the globalization of the economy and the replacement of human labour by technology in the workplace, individual purpose of life, as conceived during the industrial age, may get harder to reproduce in the future. The technologically driven world is quickly side-lining labour in the production process producing great anxiety among a large portion of the population who have made their living in the manufacturing economy. The service industries are not immune to this type of alteration in the technological world, either. With the reduction of labour in the production process, it is not clear that work in the new technological age will be sufficient to replace what is lost through the reduction of workers in formerly labour-intensive manufacturing. The same can be said about much of the service sector where a great amount of administrative work now is being done by machines. This loss includes not only income but also the avenue for maintaining self-esteem and life meaning.

The transformation of society through the generation and adoption of a new organizational framework and social contract will drastically affect the way we relate to one another and the physical environment on which life depends. Like previous human transformations, it will require a reorganization of individual and community values on which the social superstructure is constructed. The need for this transformation is evidenced by the increasing political and economic unrest we witness today and the rise in nationalism spreading throughout many parts of the globe. The fear and anger felt deeply by many citizens is now bubbling to the surface in numerous countries and manifests itself in what I see as destructive and unproductive actions – Brexit, for example. The rise and increasing strength of ultra-right-wing organizations such as the UK Independence Party (UKIP), which was instrumental in carrying the banner for Brexit (Britain's exit from the European Union), and similar organizations in many other European countries, is driving the clamour for referenda with the intention of leaving the European Union. The underlying motivation for many of these ultra-right-wing parties is xenophobia and anti-immigration. Also, a great deal of this anger is directed at governments and politicians who are thought to be deaf to the problems of their constituents.

Nationalism is often a sign that a country is faltering and in decline. It appears that the resurgence of nationalism is an attempt to gain something that has been lost. There is no doubt that many citizens in the hegemonic countries have lost their privileged position and some have become marginalized in society. The problem is not the amount of wealth generated in these countries but its inequitable distribution among the population. Unfortunately, the wealthy believe that by becoming nationalistic, they can blame others for their greed and the unfair trade deals that negatively affected many people. They do this in hopes that they will escape being identified as the actual cause. It is an old game of distraction: divert attention away from the real location of a country's internal problems.

It is often the case that xenophobia masks the deeper problem of disenfranchisement and loss of social position by a large group of formerly privileged citizens. When people feel they are being left behind economically by a society that has promised so much, they often succumb to anti-government and anti-immigration sentiments when those promises are not realized. There needs to be someone or something to blame for one's social pain and disenfranchisement. The immigrant is this generation's target, although immigration has always generated some anxiety among the population in western democracies. There appears to be even less tolerance for the 'other' today than in times past.

The basic macrosocial system in which we live today is a global interaction and exchange system; I say this with some reservation. The economic system is truly global, and we have a partial global political system – in fact, we have several of them. For example, there is the European Union (EU); the Association of Southeast Asian Nations (ASEAN); and the United States, Mexico, Canada Agreement, formerly the North American Free Trade Agreement (NAFTA). In addition to these trading partnerships, there are other global quasi-governmental organizations, such as the United Nations (UN), the World Trade Organization (WTO), the World Bank (WB), and the International Monetary Fund (IMF). Beyond these financially oriented institutions, there is also the World Health Organization (WHO) with a mandate to monitor and respond to health crises as they emerge throughout the world. Additionally, their mandate includes increasing the level of health of populations in the less-developed parts of the globe. These systems have become more complex and entangled over time without a political system with enough power to operate them efficiently and effectively. Notwithstanding the recent retreat of Britain from the European Union, I suspect that full integration will constitute the way forward for humanity in the future rather than retrenching into state isolationism, even though globalization is under attack at the moment. No doubt the same tensions were evident when society moved from the city states to the nation state centuries ago. Today, cultures flow into one another and overlap unlike previous times when there were strict boundaries and great space separating them. Rarely did they collide. Modern forms of communication, including air travel and social media, have put an end to the isolation of cultures. We are truly a global village in today's world.

This book is built on the premise that a new social organizing system and social contract is needed to meet the latest challenges presented by the 21st century. In fact, a new world-system was probably needed in the latter part of the last century. Reformation of the present world-system is not enough to deal with the present crisis. The new technologies, the financialization of social relations, and the sophisticated transportation networks that dominate our lives demand further integration, not a return to a simpler world. There may be need for greater intensification of the global interaction networks and trading systems to successfully transition to the next stage of socio-cultural evolution rather than to retreat from them as some seem to want to do today. The organizational architecture that was created to address the issues of the industrial age is no longer relevant to the fresh conditions presented by the global warming crisis, an increasingly technological world, and issues of social and economic injustice. Any new social contract will need to be designed to enhance the life conditions of those at the bottom of the socio-economic order and not be solely focused on the interests of the economic elite or the major corporations, as in the present version of the globalized system created during the 19th and 20th centuries.

A good portion of this book is the combination of original thought and summarizing the historical and contemporary work of others. I have attempted to introduce classic and contemporary theories and the views of thinkers from the past as well as the present. I have done so, to build a framework for analyzing the present condition and for adapting theory to the needs of today in hopes of generating a new world-system model. In the chapters that constitute Part 1 of this book, I have often provided a summary of the expert research that has analyzed the most up-to-date and reliable information on the subject being discussed. For example, in the chapter dealing with climate change and global warming, I have relied heavily on the reports of the Intergovernmental Panel on Climate Change (IPCC) and the US Subcommittee on Global Change Research: Climate Science Special Report.

Although it is often easy to identify problems such as those discussed here, and the need for social policies to address these inadequacies, it is much more difficult to articulate the mechanisms and processes society can employ to change their way of life and solve these perilous concerns. Tinkering at the margins of the present capitalist economy will not suffice. A new social contract that reorders the world's social architecture, interaction networks, and basic value system is required today, much like it was at the beginning of earlier social transformations experienced by human society. But before jumping right in to creating potential solutions, it is imperative to identify and outline the major problems facing humanity that demand resolution.

Bibliography

Eisenstein, C. 2007. *The Ascent of Humanity: Civilization and the Human Sense of Self [Kobo version]*. Kobo.com.
Fukuyama, F. 2014. *Political Order and Political Decay: From the Industrial Revolution to the Globalization of Democracy [Kobo version]*. Kobo.com.
Giroux, H. A. 2018. *American Nightmare [Kobo version]*. Kobo.com.
Harvey, M. 2019. *Utopia in the Anthropocene: A Change Plan for a Sustainable and Equitable World [Kobo version]*. Kobo.com.
Intergovernmental Panel on Climate Change. 2014. *Climate Change 2014 Synthesis Report*. Geneva, Switzerland: Intergovernmental Panel on Climate Change.
Mayer, J. 2016. *Dark Money: The Hidden History of the Billionaires Behind the Rise of the Radical Right [Kobo version]*. Kobo.com.
McMurtry, J. 1999. *The Cancer Stage of Capitalism*. London: Pluto Press.
Reid, D. G. 2017. *Social Policy and Planning for the 21st Century: In Search of the Next Great Social Transformation*. London, UK: Routledge.
Schumpeter, J. A. 2016. *Capitalism, Socialism and Democracy, 2nd. ed. [Kobo version]*. Kobo.com.
Snyder, T. 2018. *The Road to Unfreedom [Kobo version]*. Kobo.com.
Streeck, W. 2016. *How Will Capitalism End? [Kobo version]*. Kobo.com.
Wallerstein, I. 2004. *World-Systems Analysis: An Introduction [Kobo version]*. Kobo.com.
Wallerstein, I., Collins, R., Mann, M., Derluguian, G., and Calhoun, C. 2013. *Does Capitalism Have a Future? [Kobo version]*. Kobo.com.

2
CLIMATE CHANGE
The elephant in the room

> The world has warmed more than one degree Celsius since the industrial revolution. The Paris climate agreement – nonbinding, unenforceable and already unheeded treaty signed on Earth Day in 2016 – hoped to restrict warming by two degrees. The odds of succeeding, according to a recent study based on current emission trends, are one in 20. If by some miracle we are able to limit warming to two degrees, we will only have to negotiate the extinction of the world's tropical reefs, sea level rise of several meters and the abandonment of the Persian Gulf. The climate scientist James Hanson has called two-degree warming 'a prescription for long-term disaster'. Long-term disaster is now the best-case scenario. Three-degree warming is a prescription for short-term disaster; forests in the Arctic and the loss of most coastal cities. Robert Watson, a former director of the United Nations Intergovernmental Panel on Climate Change, has argued that three-degree warming is the realistic minimum. Four degrees; Europe in permanent drought; vast areas of China, India and Bangladesh claimed by desert; Polynesia swallowed by the sea; the Colorado River thinned to a trickle; the American Southwest largely uninhabitable. The prospect of a five-degree warming has prompted some of the world's leading climate scientists to warn of the end of human civilization.
>
> (Rich, 2018, Prologue)

The forgoing warning is taken from the prologue to an article in the *New York Times* magazine in August 2018 written by Nathaniel Rich. It is a foreboding but accurate assessment of the disaster ahead if humanity does not pay attention to the warnings of the science community about the consequences of climate change and global warming. And, it doesn't seem that we are paying attention.

Eisenstein (2007) raises the issue of humanity's separation from nature, treating it as if we were not a part of it and thinking that we can control it with impunity. The human community at all stages of development has attempted with great abandon to control nature and wantonly overuse the planet's resources without regard for the consequences. In fact, we have thought there were no bounds to resource use and have completely ignored the consequences of those activities. We have fouled the water, soil, and most importantly the air by unceremoniously dumping refuse from the resources we use into the environment. Humans have taken the Christian Bible literally, and since Genesis 1:28 was written, we have attempted to 'be fruitful and increase in number; fill the earth and subdue it. Rule over the fish in the sea and the birds in the sky and over every living creature that moves on the ground'. This reckless destruction of the natural environment is now taking its revenge. I agree with Eisenstein's premise that humanity has separated itself from nature, in an attempt to control it. I don't agree with Eisenstein when he suggests that nature has a purpose and that humanity needs to determine that purpose and work with nature to achieve it; it is a comforting idea but not one that is likely to produce the desired outcome. As Eisenstein himself suggests, this is inherently teleological thinking. According to Darwinian evolutionary theory, nature does not have a purpose other than to survive and propagate. He goes on to suggest that we cannot control nature and with that I agree. What we can control is ourselves and how we approach our relationship with nature. Unlike Eisenstein's reliance on a metaphysical approach to our relationship with nature, one that downplays rationality in favour of a more spiritual concept, I argue that we must increase our rational, fact-based approach to decision making when concerned with the environment and preservation of nature. We must establish a clear goal for preserving nature, not simply maximize our use of it. Our emotional logic is enacted when we profess our love for nature, but our activities toward it must be based on scientific fact-finding and analysis. Humanity must realize that our future depends on our clear-eyed understanding of nature and symbiosis with it. Human activity must be directed at protecting and enhancing nature with the understanding that all life, including ours, depends on its health.

This chapter summarizes and interprets the climate change and global-warming literature that has been produced over the past few years by the world's leading climate scientists. As these reports are published, it becomes clear that the planet is suffering from severe climate change, leading inevitably toward global warming. In fact, global warming is now being felt by all of us every day and it is only going to get more invasive as time goes on. One of the major issues pushing humanity toward major change to the social system is the advent of climate change resulting in global warming. As the USGCRP, Climate Science Special Report: Fourth National Climate Assessment, Volume 1, states,

> (s)ince NCA3, stronger evidence has emerged for continuing, rapid, human-caused warming of the global atmosphere and ocean. This report concludes that it is *extremely likely* that human influence has been the dominant cause of the observed warming since the mid-20th century. For the warming over the last century, there is no convincing alternative explanation supported by the extent of the observational evidence.
>
> *(USGCRP, 2017, p. 12)*

That said, the scientific community is rarely, if ever, able to state any proposition with absolute 100% certainty. Science deals in probabilities; therefore, the highest certainty that is possible in any of these reports is 'extremely likely' or some equivalent phraseology.

If the world is warming because of human activity, and there seems to be very little doubt about it no matter what the climate change deniers want us to believe, society must change at its core to effectively address the pending crisis. In fact, global warming is not only pending, given the wildfires, droughts, and floods that are occurring around the globe every day, but evidence suggests it's already here. Climate change and global warming are happening now. It is becoming increasingly evident that human-made climate change is having a profound effect on the planet and it is quickly leading to environmental disaster.

We are witness today to the results of climate change and global warming. There has been an unprecedented increase in the number and intensity of wildfires on the west coast of Canada and the United States. Wildfires are also increasing their presence in Europe and other countries around the world. The Amazon forest, long thought to be the lungs of the planet, is under considerable pressure through logging and agriculture. Further, fires have been particularly devastating in Australia and other Pacific-region countries. Wildfires are no longer considered to be an anomaly but a regularly occurring event. They are beginning earlier in the season and lasting longer than in previous years. Their intensity has also increased dramatically. They consume vast amounts of forest timber, which adds to the CO_2 in the environment and eliminates trees that would otherwise provide a huge carbon sink absorbing a considerable amount of carbon from the atmosphere. The most recent assessment by 13 departments in the US government suggests:

> (t)he last few years have also seen record-breaking, climate-related weather extremes, the three warmest years on record for the globe, and continued decline in arctic sea ice. These trends are expected to continue in the future over climate (multidecadal) timescales. Significant advances have also been made in our understanding of extreme weather events and how they relate to increasing global temperatures and associated climate changes. Since 1980, the cost of extreme events for the United States has exceeded $1.1 trillion; therefore, better understanding of the frequency and severity of these events in the context of a changing climate is warranted.
>
> *(USGCRP, 2017, p. 12)*

The same increase and ferocity of catastrophic natural events is also occurring in southern Europe, particularly in Greece, Spain, and Portugal. Australia has always been subjected to wildfires, but the situation seems to be getting worse. Wildfires are a global phenomenon and a direct result of humans emitting too much carbon and methane into the atmosphere. In addition to forest fires, other changes due to climate change are also occurring. Ironically, the opposite of wildfires, catastrophic weather resulting in increased flooding is of critical concern in many parts of the world, particularly in Asia and some areas of North America. The warming of the climate has changed weather patterns, leading to heightened storms with increased catastrophic winds, tornados, hurricanes, and typhoons in all parts of the world. Rising ocean levels are threatening to inundate islands and low-lying land. Some major parts of the world's biggest cities could be threatened with permanent flooding if ocean levels continue to rise at the rate they are rising today. As the ice caps on both poles continue to melt, the Arctic land and Antarctic water that the retreat of ice exposes will absorb increasing amounts of the sun's radiation and elevate the temperature of the planet catastrophically. Ironically, we are now witnessing atmospheric warming from open water in the Arctic, radiating back into space and affecting the polar vortex that permanently sits in the stratosphere over the far north. The increased warmth radiating from the open oceans below is now forcing part of the polar vortex to move lower down over the northern hemisphere during the winter months, causing lower than normal temperatures over a larger portion of North America and for longer periods of time. The open water that allows the increased warmth back into the stratosphere is caused by melting ice in the Arctic. Whether this action will intensify as time goes on or is only a passing phenomenon is yet unknown.

Melting of the permafrost in the Arctic will also release underground methane, complicating the issue further. Flannery (2005) predicts that we are in a climatic downward spiral and it will take great effort to extricate ourselves. In a public presentation given by Amaya M. Davis, PhD, Solar System Ambassador from NASA in Seville, Spain, on January 10, 2019, the point was emphatically made that climate change cannot be stopped and global warming cannot be reversed. Some of the conditions she cited that are leading to this conclusion are rising global temperatures, warming oceans, shrinking ice sheets, glacial retreat, decrease in snow cover, rise in sea level, declining Arctic sea ice, and ocean acidification. Her conclusion is to slow it down and learn to adapt.

How long have we known about this crisis? Science has been warning us of the impending disaster since the early 1970s, perhaps even earlier. The talk of climate change in popular culture is more recent. Perhaps the most substantial and noticeable popular piece of work was Tom Flannery's book (Flannery, 2005) *The Weather Makers: How We Are Changing the Climate and What It Means for Life on Earth*. It was Flannery's book in which I first encountered a substantial personal account of climate change and global warming. He was able to interpret the scientific and technical language into popular literature. His book not only interpreted the science of global warming but also articulated his experience in a manner that could appeal to the lay reader. On the first page of the book, he portrays his initial encounter with the tangible effects of climate change during a walk in a forest in New Guinea. He cites a surprising change in the flora that indicates in real time the changes produced by climate change.

> In the leaf litter on the forest floor I was surprised to find the trunks of dead tree-ferns. Tree-ferns grew only in the grassland, so here was clear evidence that the forest was colonizing the slope from below. Judging from the

distribution of the tree-fern trunks, it had swallowed up at least thirty metres of grassland in less time than it takes for a tree-fern to rot on the damp forest floor – a decade or two at most.

(Flannery, 2005, p. 1)

He goes on to ask,

> Why was the forest expanding? As I pondered the moldering trunks, I remembered reading that New Guinea's glaciers were melting. Had the temperature on Mt. Albert Edward warmed enough to permit trees to grow where previously only grasses could take root? And, if so, was this evidence of climate change? . . . The experience left me troubled; I knew there was something wrong, but not quite what it was.
>
> *(Flannery, 2005, p. 2)*

Much has happened since the time of Flannery's writing. The world appears to have taken little notice of his and other writings about global warming and not much has been accomplished in addressing the matter despite heroic efforts by climate advocates to define the problem for the public and their governments. It would seem that the concern with the crisis of global warming cannot compete with those who put the economy and their ability to create more wealth above all other social and environmental interests. If humanity is going to come to grips with climate change and global warming, there will need to be a huge shift in culture, particularly in the affluent, capitalist economies, and this idea doesn't seem to be getting much traction at the moment.

What is particularly distressing is the lack of political will in addressing the problem, and the United States, the largest economy on the earth and supposedly the leader of the free world, has now turned its back completely on the issue. Shortly after taking office, the current president of the United States withdrew from the Paris Climate Accord and has subsequently rolled back regulations that were designed to protect the environment and climate by reducing CO_2 emissions. The United States is heading back to the 1950 standards for regulating emissions of such commodities as coal and oil production. Auto emission standards are also being severely rolled back. The present administration is a climate change denier (they may not deny that climate change is occurring; they see it as a natural phenomenon and not necessarily human made). As Wallerstein tells us,

> (t)he least expensive way for a producer to deal with waste is to cast it aside, outside its property. The least expensive way to deal with transformation of the ecology is to pretend it isn't happening. Both ways reduce the immediate costs of production. But these costs are then externalized, in the sense that either immediately or, more usually, much later, someone must pay for the negative consequences.
>
> *(Wallerstein, 2004, p. 81)*

It would appear that rather than having polluters underwrite the cost of clean-up, society is about to pay for the long-term externalization of pollution through global warming and its negative consequences. Ironically, the public will likely pay a lot more in the end than if we had acted earlier.

Lobbyists and friends of the fossil fuel industry have taken over the highest levels of the Environmental Protection Agency (EPA) in the United States. They are now in the process of dismantling regulations, inadequate as they have been, that were designed over the years to protect the environment, including addressing the increasing temperature of the earth. Auto emissions have been rolled back in the name of economic growth. Economic growth is by far the major concern of most governments. Even fighting the COVID-19 virus is overshadowed by maintaining the economy for a large portion of the US population if not across the world. Today's society is still firmly planted in the capitalist system where the government focuses overwhelmingly on economic issues. As James McCarville so famously said when heading up President Bill Clinton's presidential campaign, 'it's about the economy stupid'. Everything in politics is about the economy. Our politicians can't seem to get past it and address other highly important matters no matter how critical they seem. As Wallerstein notes, '(w)e are in a capitalist system only when the system gives priority to the endless accumulation of capital' (Wallerstein, 2004, p. 45). There is no evidence that we have moved away from world capitalism or that we are even considering it in order to address such alarming issues as global warming. It would appear we are unable to act pluralistically and address issues that have long-term consequences to our species and beyond. For a brief period in the Obama era, it appeared that the United States was going to show leadership on the environmental front, specifically on the climate change file. Obama was a major supporter of the Paris Climate Accord in addition to developing many regulations under the Environmental Protection Agency to reduce CO_2 and other environmental contaminants. But, alas, the present US administration has rolled back many, if not all, of those green initiatives in short order. The present US administration has even attempted to kick-start the coal industry, perhaps the greatest emitter of CO_2. Long-standing regulation restricting oil and gas exploration in highly environmentally sensitive areas has been rolled back along with most regulations that were designed to combat climate change. These anti-environmental actions come at a time when the world is watching the collapse of the Arctic environment and the melting of the polar ice caps.

The present situation regarding climate change and the warming of the earth's surface has been researched and made known to the world and its leaders for a long time. As Wallerstein argues in his book *World-Systems Analysis*, problems such as global warming must be tackled cooperatively and internationally. The problem is not the domain of any single nation state, but the largest and wealthiest nations like the United States must provide leadership if the world is to address the issue adequately. Without the most powerful nation in the world giving direction to combating the global warming crisis, there is likely to be little progress made by the rest of the world even if the other countries act cooperatively and in unison.

In my own country of Canada, we continue to have an ambivalent notion about climate change. On the one hand, the government has introduced a carbon tax, but on the other, it has bought a failing oil pipeline that is dedicated to transporting oil to the west coast in order to export the product to the rest of the world, primarily Asia. The citizens of Canada are aware and appreciate the dangers inherent in global warming due to climate change. Even so, the government is encountering deep resistance when taking measures such as the institution of a price on pollution – a carbon tax – as they try unsuccessfully to meet their commitments under the Paris Climate Accord. A carbon tax has become law and was implemented in January 2019. Some backward-thinking provincial governments are fighting this measure in court, so we will see the eventual outcome of this policy in due course.

Scientists have been sounding the climate change alarm since the early 1970s, well over 40 years. Given the dire consequences of this phenomenon, why has humanity not acted? Nathaniel Rich asks this very question, and his answer is chilling:

> Everyone knew – and we still know. We know that the transformations of our planet, which will come suddenly and gradually, will reconfigure the political world order. . . . Could it have been any other way? In the late 1970s, a small group of philosophers, economists and political scientists began to debate, largely among themselves, whether a human solution to this human problem was even possible. They did not trouble themselves about the details of warming, taking the worst-case scenario as a given. They asked instead whether humankind, when presented with this particular existential crisis, was willing to prevent it.
>
> *(Rich, 2018, Epilogue)*

Our lack of action on this question over the last half century is a direct answer to the question posed by Rich. Is global warming an issue that the population of the world is unwilling to address before it becomes a catastrophe? This is yet an unanswered question, but given the current political climate, hope is scarce. Rich notes that 'human beings, whether in global organizations, democracies, industries, political parties or as individuals, are incapable of sacrificing present convenience to forestall a penalty imposed on future generations' (Rich, 2018, Epilogue). Until this underlying apathetic attitude of society changes, a satisfactory conclusion to global warming is unlikely. Addressing this issue adequately requires a new way of seeing the world and the development of a set of appropriate values on which a new philosophical, political, social system, and particularly the economy, can be reconstructed. Given the lack of concern on the part of the public and the United States' recent deregulating actions, meaningful change to global warming will require a reinvigorated public effort. The longer the mitigation measures are put off, the more drastic those measures will need to be to adequately address the issue. The path to mitigation must begin with a summary of what is known about the situation and the consequences of non-action. That information would then need to be presented to the public in the form of a massive educational program.

Until the public is fully engaged in this matter and mounts a pressure campaign on their governments, little progress is likely to be made. It is clear that political leadership on this file is lacking or non-existent altogether, so the drive for action will need to come from civil society. The rudiments of this effort are now under way by concerned individuals and groups in society, particularly by millennials who will face the consequences of this pending catastrophe for most of their adult lives.

The remainder of the chapter is devoted to assembling and synthesizing the major reports that have been produced over the last few years that outline the state of the changing climate and its effects on the planet and the human condition. This is one small step in creating the public educational program that needs to be mounted in order to gain public support for taking the drastic measures needed to defeat the concern. The core arguments for recognizing the reality of climate change and for initiating a public education campaign are as follows: the present increase in catastrophic weather, what the future may have in store if global warming is not checked, and what processes and actions require implementation to begin to address the issue.

Given the overwhelming evidence for climate change and global warming that has been in the literature for some time, I do not intend to re-present in detail those data or the arguments made by the IPCC or the many other reliable reports that present a similar picture. I will, however, summarize their main findings and projections. Perhaps what is most important to do here is to discuss what needs to be done in the face of the scientific evidence for climate change. In some respects, discovering the evidence for climate change and global warming is relatively easy compared to determining what needs to be done about it and how to go about doing it.

Climate change is a subject that was much discussed and debated in society during the last three decades of the 20th century and the first two of the 21st century. While that discussion continues today, insufficient concrete action has been taken to adequately address the problem to date. The physical evidence for climate change has been well documented particularly by the Intergovernmental Panel on Climate Change (IPCC) in their many reports. The IPCC is a body of recognized scientists from many parts of the world and legitimized by the United Nations. Their work and reports have achieved legitimacy throughout the world even though some climate deniers attempt to undermine their credibility. The IPCC is made up of thousands of credible scientists who have expertise on climate change and related matters. Some of this chapter will present their main arguments to set the context and demonstrate the significance of global warming and remind us of what we already know about the subject. More importantly, this chapter addresses the political, economic, and social implications of global warming as it relates to everyday life.

One of the first substantial studies produced on climate change in the United States was conducted by the Board on Atmospheric Sciences and Climate; the Commission on Physical Sciences, Mathematics, and Resources; and the National Research Council. Their 1983 report (Board on Atmospheric Sciences and Climate, 1983) was titled *Climate Change: Report of the Carbon Dioxide Assessment*

Committee. Although they admit in their report that they had a few uncertainties about the ultimate effects of climate change, they were convinced that if the trends they identified were to continue unabated into the future, climate change and global warming would be an unalterable fact. More recent studies have confirmed their suspicions emphatically, and we see some of their dire predictions playing out on the ground today. In addition to confirming some of their suspicions about climate change and global warming, new estimates have been added to their concerns and revised approximations about the speed with which some of their major predictions would materialize. A more recent and authoritative report on the issue was completed by the IPCC, who not only confirmed the fears of the authors of the 1983 report but have also increased the estimates of the consequences of global warming on the natural systems of the planet and the speed with which those consequences will occur. The updated report suggests there is greater urgency for addressing the issue of climate change and global warming today than when the 1983 report was published. The report advises that unless society addresses this issue assertively and keeps climate change to an increase of only 1.5°C, we will pass the point of no return within the next 15 years and lose our ability to head off drastic changes to the climate and warming of the planet. Although the IPCC provides this dire warning, the sitting president of the United States, in response to a question posed to him after their 2018 report was released commented, that "the climate is fabulous." This response should not provide humanity with confidence that the issue is understood or even recognized, let alone being addressed in a positive manner by the most powerful and influential country in the world today. In spite of the total neglect of the issue by the US government, many individual states, municipalities, and some corporations are instituting measures to at least give the issue some recognition, but their efforts are totally insufficient.

The IPCC confirms that the world has warmed in every part of the globe. The annual average surface air temperature has warmed by about 1.6°F (0.9°C) over the last 150 years. This warming has triggered many other changes to the earth's climate. Evidence for a changing climate abounds, from the top of the atmosphere to the depths of the oceans. Thousands of studies have documented changes in surface, atmospheric, and oceanic temperatures. We are witnessing melting glaciers, disappearing snow cover, shrinking sea ice, rising sea levels, and an increase in atmospheric water vapour. Evidence demonstrates that human activities, especially emissions of greenhouse (heat-trapping) gases are primarily responsible for recent observed climate changes. North America has experienced increases in forest fires, floods, droughts and many more tornadoes, hurricanes, and severe storms than is normal over recent history. Some parts of the world are experiencing record-breaking temperatures on an ongoing basis.

It is understood that the great culprit in climate change and global warming is the use of fossil fuels at an accelerating rate to increase energy production. The most urgent actions required to address this problem is to reduce the use of fossil fuels. CO_2 is the major culprit in producing greenhouse gases that create a layer in the upper atmosphere. It blocks the sun's heat from escaping back into space after landing on earth, thereby warming the earth's temperature. Essentially, it acts like a greenhouse.

Ground zero in the fight against climate change is the stabilization of temperature increases to below 2°C relative to pre-industrial levels. As mentioned earlier in this chapter, the new 2018 report of the IPCC has revised that number down to 1.5°C. No doubt this may be very hard to do. The IPCC warn us that the longer we wait to deal with climate change, the more difficult it will be to avoid a catastrophe. They are emphatic that climate change, more specifically global warming, is a consequence of human economic activity. It is an inarguable fact that the climate has warmed considerably since the advent of the Industrial Revolution.

While humans have become exceedingly proficient in utilizing the world's natural resources, we have not paid enough attention to the environmental consequences of dumping waste from the use of those resources onto the land and into the water and air. And, while some regions of the globe have benefited directly by industrialization, other areas have not profited as much. Many have been substantially left out of economic development and the benefits it has provided. This complicates the picture considerably when the reduction of CO_2 discussion is held. Many of the developing countries feel, and rightfully so, that they are now being asked to pay for the reduction of global warming when they have not been a primary participant in creating it. Regardless of this historical fact, global warming is not area specific but proliferates throughout the globe and affects polluters and non-polluters equally. As the 2017 US Climate Change Report notes:

> Understanding the full scope of human impacts on climate requires a global focus because of the interconnected nature of the climate system. For example, the climate of the Arctic and the climate of the continental United States are strongly connected through atmospheric-circulation patterns. While the Arctic may seem physically remote to those living in other regions of the planet, the climatic effects of perturbations to Arctic sea ice, land ice, surface temperature, snow cover, and permafrost affect the amount of warming, sea level change, carbon cycle impacts, and potentially even weather patterns in the lower 48 states. The Arctic is warming at a rate approximately twice as fast as the global average and, if it continues to warm at the same rate, Septembers will be nearly ice-free in the Arctic Ocean sometime between now and the 2040s.
>
> *(U.S. Global Change 2 Research Program, 2017, p. 14)*

Although the affluent industrialized countries of the world, including India and China, have contributed the greatest amount of greenhouse gases to the atmosphere, leading to climate change, all countries of the world are now suffering the consequences of that past behaviour. In addition to the factual changes to climate and weather patterns around the world we witness today, the process of global warming seems to be increasing each day. To date we have not been able to stop its progression. We are still adding CO_2 to the atmosphere each year, not reducing it. It is important to determine where we are today regarding the issue of climate change and the likely progression of events if nothing is done to reduce or limit global warming completely.

What is the present state of climate change and global warming to which all regions of the world are affected? To begin this discussion, it is worth laying out what I interpret to be the most fundamental conclusions of the IPCC (2014) and the U.S. Global Change 2 Research Program (2017).

Table 2.1 provides a snapshot of where we are today in terms of climate change and global warming and our role in producing it. It is a fast-changing picture and the speed with which these estimates are becoming reality is surprising even to the scientists contributing to the many climate change studies that have been produced to date. What are the direct changes that affect our lives now and will continue to do so into the future? Here are just a few of the consequences of climate change we should be concerned about. Again, much of this information is gleaned from the IPCC report of 2014.

Anthropogenic greenhouse gas emissions have dramatically increased the concentrations of carbon dioxide, methane, and nitrous oxide in the atmosphere since industrialization. This, of course, has led to global warming. As a result, the ice caps at both ends of the planet are melting. This will add great amounts of water to the oceans around the world. The complete melting of the ice on Greenland, for example, will result in a 20-foot rise in sea levels. The IPCC estimates that the melt rate is in the range of 3.5% to 4.1% per decade. This rise in sea level will dramatically impact the inhabitants of islands and coastal cities around the world. Many islands will disappear completely, and coastal cities will be repeatedly inundated with water far into the future. A large rise in sea level may also affect ocean

TABLE 2.1 Fundamental conclusions on global warming

1 The major portion of CO_2 and methane emissions emitted into the atmosphere is a result of human activity.
2 The human community has been unable to reduce total emissions into the atmosphere to date. In fact, we are still adding to the CO_2 content of the planet each year.
3 The increase in CO_2 is not region specific but felt across the entire globe.
4 The warming of the planet is demonstrable and has been since the 1950s.
5 The intensity of weather events that we are now experiencing throughout the world is directly related to global warming.
6 The warming of the globe is now an empirical fact, and what we can hope for is to limit its impact but not reverse its course.
7 Regardless of who the major contributors to global warming have been historically, all parts of the globe will need to contribute to CO_2 reduction if global warming is to be controlled.
8 As the ice shields in the Arctic and Antarctic melt at an increasingly quicker rate, coastal areas will continue to flood, including some major seaboard cities around the globe.
9 The speed with which global warming is occurring is increasing and exceeding the estimates of the scientific community.

circulation around the globe, which, in turn, will produce extreme weather events throughout all regions of the world. Ocean warming and the acidification of the seas have led to the decline of fish species and coral reefs. The amount of oxygen in the oceans is diminishing. Increasing sea surface temperatures, rising sea levels, and changing patterns of precipitation, winds, nutrients, and ocean circulation are all contributing to overall declining oxygen concentrations in ocean and coastal waters (Intergovernmental Panel on Climate Change, 2014). The rise in sea levels during the last decade has been dramatic.

The Great Barrier Reef off the coast of Australia is in great peril. Many coastal areas and inland lakes are virtually dead because of warming and lack of oxygen. An additional reason for the lack of oxygen in lakes is the fertilizer that runs off the land into rivers that feed into lakes. This process is due to intensive factory farming and increasing use of chemicals in agricultural practices. Many previously forested areas around the globe are being turned into agricultural land.

Recent extreme heat waves, droughts, floods, cyclones, and wildfires are a direct result of climate change. These weather events expose some ecosystems and many human systems to destruction and disaster. Perhaps one of the extreme changes in the north, besides the melting of the polar caps, is the thawing of the permafrost. This will have huge, negative, effects for both human communities and wildlife. The polar bear, for example, is clearly under stress and in severe decline, perhaps predicting their eventual extinction, and this is not the only species under threat. The sea ice and, therefore, the hunting areas of the polar bear are diminishing in size every year causing them to become nuisance bears as they search for food in Inuit communities. Once the permafrost thaws, previously frozen organic matter will also melt intensifying the greenhouse gas emissions exacerbating the problem even further. Great amounts of methane are locked into the ground under the permafrost and will be released into the atmosphere as the permafrost melts. This too, will accelerate global warming. In addition to the various consequences to the environment outlined by the IPCC in their many reports, the rapid melting of the permafrost in the Arctic is now alarming the climate scientists. To date, melting of the permafrost has not been modelled in the climate predictions of the science community and is occurring much faster than scientists had assumed and with greater consequence than originally thought. A good portion of the surface area in the Arctic is now covered by water in the summer as the permafrost melts. Vast portions of land are now water gardens of floating flora.

As the permafrost melts the human infrastructure in the Arctic is literally sinking into the ground, or what used to be ground. The water and sewage pipes that were laid on or in the permafrost were thought to be encased in something akin to cement. But as the permafrost thaws, the pipes are moving and breaking apart. This will require vast amounts of dollars to repair, and how long that repair will last if the permafrost continues to melt is anyone's guess. This battle is likely to be an ongoing struggle. The same crumbling is occurring to the infrastructure located above ground. The foundations of buildings are shifting and cracking causing those structures to become unsafe. This too will require huge amounts of mitigation and reconstruction dollars.

In addition to the many consequences permafrost melting places on the human infrastructure, it is also estimated that this phenomenon will release huge amounts of CO_2 and methane into the atmosphere. It is estimated that the Arctic contains twice as much carbon and methane as is in the atmosphere today. The Arctic has absorbed huge amounts of carbon since the beginning of life on the planet, and much of that carbon could be released as the permafrost begins to melt. Just how much of that carbon will be released into the atmosphere depends on how deep into the permafrost the melt goes down. Additionally, it is speculated by scientists that the permafrost may contain some ancient viruses that are unknown to humankind. It is not known what effect they will have on the human population once released into the atmosphere. Let there be no doubt, permafrost melting is a significant development. This event could intensify and hasten the already dire predictions for global warming produced by the IPCC and other scientific forums in addition to causing great havoc to the people that live in the far north.

Ice loss in the Antarctic is also prevalent. A 5,800-square kilometre piece of ice has broken away from the Larsen Shield, the fourth largest ice shelf in the Antarctic. This newly created iceberg is the size of Luxemburg. This huge mass of ice will eventually float out to sea and melt adding to the rise in sea level. One must presume that this is the first dramatic start of thawing in the Antarctic in addition to what we see in the Arctic. The rise in sea level is occurring faster than most of the original forecasts made by the scientific community, including the IPCC who have shortened the original time frame for reaching intolerable levels of warming.

As important as it is to continue to present these facts about the consequences of climate change there is nothing new in them. We have known where we are heading on this issue since the late 1980s, certainly long enough for us to have altered the course or at least mitigated it somewhat. But as Rich laments:

> Is it a comfort or a curse, the knowledge that we could have avoided all this? Because in the decade that ran from 1979 to 1989, we had an excellent opportunity to solve the climate crisis. The world's major powers came within several signatures of endorsing a binding, global framework to reduce carbon emissions – far closer than we've come since. During those years, the conditions for success could not have been more favorable. The obstacles we blame for our current inaction had yet to emerge. Almost nothing stood in our way – nothing except ourselves. . . . If the world had adopted the proposal widely endorsed at the end of the 1980's – a freezing of carbon emissions, with a reduction of 20 percent by 2005 – warming could have been held to less than 1.5 degrees.
>
> *(Rich, 2018, Epilogue)*

As we all know, that did not happen, and the consequences are seen and felt today.

The science of climate change and global warming is sophisticated and well documented. James Hanson has been a leading scientist on climate change and he consistently sounds the alarm. He is adjunct professor at Columbia University where he is the director of the Earth Institute's Program on Climate Science, Awareness and Solutions. He is one of the world's most respected climate scientists. In a recent paper Hanson suggests that we would need to not only reduce emissions but to levels below what we are putting into the atmosphere at present (see Rich, 2018). If left unattended, the warming of the atmosphere will melt the ice caps at both poles leading to the rise in the oceans and consequently the flooding of great expanses of land including cities and communities. It is not being over dramatic to suggest that whole islands will be inundated with water and in many cases covering the entire landmass. Additionally, weather patterns will become altered leading to more violent storms and great drought. This change is not some distant event but is being experienced now in many parts of the world including Europe and North America. Floods such as the one experienced due to hurricane Katrina in New Orleans in 2005; the raging wildfires, such as those which gutted Fort McMurray in Alberta, Canada, in 2016; and the melting of the permafrost in the far north of the globe are just some of the recent events that highlight the consequences of climate change.

These and many other weather-related events are causing havoc around the world and this condition is likely to get worse unless global warming is slowed down. Even if policy makers take drastic decisions and dramatically reduce CO_2 emissions this carnage will continue for decades. Turning the trend of climate change and global warming around will take years to achieve. It is like turning an ocean liner around in the middle of the sea it is not something you do with instant results. That said, it is time to act. It will become even harder to slow down the advance of global warming the longer we wait to engage the problem.

In addition to the single events triggering climate change described earlier, there are multiple events that can lead to different types of consequences. For example, wildfires or severe drought that denudes the flora of a large area may eventually produce large-scale mudslides when torrential rains and floods occur. In fact, this phenomenon is occurring in some parts of California and elsewhere throughout the world now. Given that the planet is a living system, one can only speculate that even the weather we experience as isolated singular events are in fact related to one another and perhaps many of them are indicative of tipping points in the global warming phenomenon. The sum total of the social and environmental destruction from these compound events is likely to be much greater than single events as devastating as they may be in and of themselves (Intergovernmental Panel on Climate Change, 2014). It is clear to most observers that we are now witness to unnatural climate variability. What is going on today with wildfires, drought, extreme weather (such as the polar vortex), and floods should be viewed as the new normal.

In addition to the basic alteration of the physical planet, there are also some very dire effects to individual life. Health is a major issue associated with climate change. Waterborne diseases will surge as flooding increases in intensity and duration. Respiratory conditions will increase because of hot temperatures and as airborne particulates intensify. As mentioned earlier, ancient viruses are ready to be released into the atmosphere once the permafrost in the Arctic melts. Food will also be of concern as the fertility of land changes. Catastrophic weather patterns and disastrous storms will wreak havoc on populations worldwide – damage not only to property and life itself but also on our ability to produce food as productive areas are made unproductive through drought and desertification.

We already see signs of failed crops because of weather conditions produced by global warming. Crop failure is not occurring for just a single year but repeating year after year, causing famines and widespread hunger in some parts of the world. Unfortunately, these famines are occurring in already destitute areas, mostly in Africa. Crops themselves are becoming less nourishing given that climate change is also making the land less fertile. Global warming is having great effect on the fertility of the soil and the number of crops that are being grown worldwide. It is also reducing the yield of many different varieties of crops. Global warming is certainly having a pronounced detrimental effect on agriculture worldwide and it may continue to get worse as the earth continues to warm. On the positive side, new areas of the world that were not suitable for agriculture in the past will undoubtedly open up to new crop production. As time goes on and the globe continues to heat up, this increase in newly productive land will not offset the loss of productive territory, however.

Climate change is not only a scientific fact-based phenomenon but also a political issue particularly when it comes to mitigation measures. In fact, global warming is a political problem as much as it is a scientific one. Those who choose not to accept the legitimacy of climate change, particularly the propagation of it by humans, do so because of some personal agenda or basic ideology that they wish to preserve, or the unfettered freedom to act as they please at the expense of the environment and, perhaps, life itself. There is some perverse notion among many climate deniers that to accept the science would in some way limit their basic right to political freedom. This notion needs to be discredited. The facts show that to allow climate change to continue unaltered will restrict future freedom, both political and behavioural, and, perhaps more importantly, the freedom produced by life itself.

One of the greatest fears when dealing with climate change is the conventional wisdom that to address it effectively, society, particularly affluent society, will need to reduce its standard of living. Most advocates of climate change mitigation argue that reducing the standard of living substantially may not need to be the case. They assure us that society can address global warming and continue to grow the economy at present or at accelerated levels. Is this in fact the case? Problematic is the notion that society will be required to sacrifice their economy and accept a lower standard of living than they enjoy now if this problem is to be defeated. While there are many issues and questions contained in this assertion, the basic premise, that the economy needs to be sacrificed to address the issue of global warming, may be more myth than fact. William Nordhaus, 2018 Nobel Laureate, was instrumental in creating the Dynamic Integrated Climate-Economy model known as DICE that includes climate change in its calculations. It treats the climate system as capital stock where other models ignore it completely in their calculations (Nordhaus, 2017). To treat the CO_2 pollution of the land, air, and water as a non-economic event is not a realistic proposition as we experience the economic effects of wildfires, floods, droughts, hurricanes, and so forth. The cost to the economy of these events is becoming more dramatic each year. Nordhaus argues that we can continue to grow the economy while addressing this issue. Only time will give us the answer to that dilemma; however, the case against losing ground economically if we tackle the climate problem immediately is mounting every day. Regardless as to whether addressing global warming is an economic cost or benefit, the human community must make mitigation a priority over maximizing economic growth. If global warming is not halted, there will eventually be no economy.

Climate change and global warming will open up many new industries. Green technology is a major economic sector that is now just beginning to be explored. Solar panels that collect the sun's rays are now making a substantial contribution to electrical grids in many nations. Wind turbines are also contributing to the energy supply. Tidal and wave power are also in the mix. Nuclear power has been contributing to our energy needs for some time but that too may be a mixed blessing.

Countries that have accepted the inevitability of global warming are in the forefront of manufacturing the new green technology. Retrofitting homes and commercial buildings to be more energy efficient can also have a major impact on reducing CO_2 emissions and have a positive impact on the economy. Green technology can reduce the costs of rebuilding after damage to life and property if we can bring global warming under control. It can forestall costs that have not yet occurred but will if we leave the issue unattended and left to continue its present course. So, the answer to the economic question is that it may cost us more to address the consequences of global warming if we leave it to continue its present path unattended than it would if we addressed the issue now. We should at least slow it down and, hopefully, stave off the most drastic predictions if left to carry on undeterred.

Critical to resolving this planetary predicament is the creation of a new social contract that squarely faces up to global warming and other major social and environmental issues. This is no mean task. It will require nothing less than a great leap in collective imagination not only by the leaders in society but also by all of us. This is most urgent in a time when the US government is filled with climate change sceptics and has withdrawn from the Paris Climate Accord and is aggressively eliminating environmental regulations put in place over previous years. The change in the US attitude toward climate change is forcing other governments to reconsider their CO_2 reduction strategies in order to remain economically competitive with the United States. Canada has introduced a carbon tax but may reconsider that action in light of the US position. Canada is not likely to meet its Paris Accord targets – if the carbon tax initiative is cut back or eliminated altogether, this outcome will certainly be the case. It can be expected that other governments will follow suit and give competitive economic reasons for their reduced efforts. It must be pointed out, however, that scientists are the most sceptical people on earth and there is very little climate change scepticism among them. The majority of climate scientists accept the premise that a major portion of global warming is the direct result of human industrial pollution and the overuse of fossil fuels in producing energy.

Although climate change and global warming have been clearly demonstrated by climate science, and the conclusions of the IPCC are agreed to by an overwhelming number of members of the scientific community, there remain those among us who deny its existence in the face of the overwhelming evidence to the contrary. This should not be surprising given that a significant number of the population still has difficulty accepting the role of evolution in human development. The IPCC has presented overwhelming evidence that leads to the conclusion that humans are the main cause of changing the world's climate. The cause of climate change is not in doubt. What to do about it and the sacrifices needed to be made by society is where the focus for debate should be placed.

Can humans save the planet from the scenario that predicts increases in temperature to be between 2°C and 4°C? This may be the most consequential question ever asked. The continuation of the human species, if not all species, on this planet rests on an affirmative response to this question. Addressing the issue adequately will require political will and new technological solutions. These alone, however, will not be enough; an all-encompassing solution will be required. This is one social problem that cannot be left entirely to the politicians; it will take all of us to subdue this drastic reality. A new macro approach to everyday life must be created through concerted demands by the public forcing the political system to examine and address grand questions such as global warming from a fundamentally different perspective, even though it may not have any immediate political payoff. This perspective will lessen the emphasis of the economic advantage or disadvantage of any action and concern itself with larger values of planetary health and survival. The new approach must be bipartisan and not an ideological issue. All societies across the world will need to collectively and cooperatively address the problem.

Second, policies and programmatic approaches must be global and not country specific. It is not useful for one country to embark on reducing emissions while other countries increase their CO_2 output. In this regard, the notion that some countries should be exempt from any emission regulations because they are developing and because they did not contribute to the problem in the first instance must be dismissed. Although the premise may be true, this type of exemption will only detract from addressing the problem adequately. Certainly, the affluent countries can pay restitution for their previous negligence, but all countries must contribute to rectify this overwhelming problem in one way or another if there is to be any measure of success achieved. And success means that humanity gets the problem under control and survives and not goes the proverbial way of the dinosaur.

Third, the problem of climate change and global warming and its eradication is fundamentally a moral and not an economic question. No longer can solutions to this problem be delayed because it may be detrimental to the economy of one or several nations. The fundamental economic problem of scarcity has been solved and the human community under capitalism has turned its attention to accumulation of wealth – in some cases, extreme and excessive wealth, not simply adequate and fair existence. Granted that some countries and individuals still suffer from poverty and depravation as outlined in Chapter 4 in this book, but the issue is not a matter of scarcity of goods and services but a matter of properly distributing the wealth created in an equitable manner. Humanity must triage its most pressing issues and solving global warming must take precedence over economic growth if humanity is to survive on this planet. Certainly, society can multitask and address several issues at the same time, but the issue of warming of the planet must take precedence. If global warming is not mitigated, solving the rest won't matter. Global warming and the environmental issues related to it must be placed at the head of the emergency list.

Many of the remedies useful in addressing global warming are known to us. Such public policies as a carbon tax have been suggested and implemented successfully in some jurisdictions. Incentives for the great polluters to reduce their carbon footprint, such as cap and trade, are available to policy makers in addition to a carbon tax. Advances in green energy production is gaining momentum as this book is being written. Solar, tidal, and wind power are growing in feasibility and popularity and becoming more economic as they are integrated into the electrical energy grids across the globe. Autos however begrudgingly, are moving to electrical power and away from carbon-based fuel. Hopefully, public transit will also play a heightened role in the near future. What is important to stress here is that the old technologies will continue to be refined and new ones developed to help in the cause of containing the growth of the earth's temperature. Regardless of these initiatives, humanity must resist relying on the notion that we don't need to reduce and conserve; we can't wait for a technological fix. It must be stated emphatically that there is no single silver technological bullet on the horizon that will save the planet at the last minute and there is not likely to be one ever; even if there was ever to be one, we can't wait for it to arrive in time to save the planet in case it never comes. It would be like 'waiting for Godot'. The climate predicament must be viewed as a moral, political, and power struggle, and we need to address it from that perspective and not rely on an easy but non-existent technological remedy. Also, there is no time to wait; many prominent scientists like the late Stephen Hawking believe that we may have passed the point of no return already. As eminent and learned as he was, let's hope he was wrong on that score.

Here is a list of actions that should be embarked upon immediately to get the ball rolling: retrofitting buildings to make them more energy efficient; increasing public transit to get people out of their cars; electrification of autos and the continued increase of meters (miles) per gallon (litres) of petrol; funding for the transformation of industrial and municipal waste to green energy; the introduction of a pollution or carbon tax (or other types of schemes such as cap and trade) as a disincentive to emit carbon into the air by large polluters and replace carbon-based energy with wind, solar, and hydro (tide) energy production; and, perhaps most importantly, phase out CO_2-producing industries in carbon-based petroleum industry areas and introduce schemes such as the guaranteed basic income, or other such measures, in order to lessen the effect of job loss in the CO_2-producing sectors of the economy. While these programs alone will not get us to where we need to limit global warming to only 1.5°C, it will start the process and increase visibility for the urgency to act.

What is needed is political will to act boldly. Unfortunately, there doesn't seem to be much political will around right now for such a project. As a consequence of the lack of will to act on the part of politicians and the public at large, there is need for a comprehensive public education and marketing campaign to alert the public to the emergency at hand, if they are not already informed through the recent reports by the IPCC and the renderings of other environmental scientists and advocates. This campaign must be led vigorously by leaders of society. In conjunction with a massive public educational campaign that outlines the consequences of global warming and what needs to be done about it, we must restore the credibility of our scientists and experts. Their credibility has been systematically and deliberately destroyed by those in power over the last few years.

As stated earlier, the warming of the planet is speeding up and may be approaching the point of no return, so immediate action is required. The matter is complicated by the election of a government in the United States which questions the legitimacy of the concept of global warming and the human contribution to it. Although it is likely the present US administration will accept the legitimacy of the concept in the end, it may require too much time for them to arrive at that conclusion in time to act effectively. Although much of the focus on the world's inability to successfully address global warming has been pinned on the negative actions of the United States, the large countries of Asia, including India and China, must share much of the culpability for the continued progression of global warming. These countries are still attempting to meet their growing needs for energy through coal-fired generation. Construction of coal-fired energy-production facilities continues in China today. No matter what other countries, including the United States, do in relation to CO_2 reduction, the problem of global warming will not be resolved without the full participation of the large Asian countries. Governments throughout the world have not been able to agree that climate change is real and global warming the result. Fortunately, more and more of the holdouts are beginning to realize global warming is real, as the number of catastrophes in their jurisdictions rise and the link to climate change becomes unmistakable. Time is of the essence, however.

Thus, the world now faces a life-threatening environmental catastrophe produced by our past success. The major question to be addressed is, can we move away from an economic system where a small cadre of elite individuals and special interest groups hold overwhelming political power and control over the economic and political system and move to something that will challenge the status quo embraced by the plutocracy? You can bet that those at the top of the present hierarchy will not give up their position in society and their vast wealth easily, even in the face of perpetuating great harm to individual life and society generally. This fact is very much in evidence when you observe how vehement these individuals deny climate change and defend the primacy of the market in human relations. They go to great lengths to construct hegemony over the ideas and myths that guide the social system through such actions as pouring vast amounts of money into political campaigns to have their surrogates elected to the highest offices in government. Notice how the National Rifle Association (NRA) controls the gun debate in the United States. Congress, despite the will of the majority of citizens who favour some form of gun control, continues to deny the culpability of guns in their country's astronomical homicide rate. The power structure continues to protect the gun lobby and their interests. The same can also be said on the climate change front.

There are a number of policy and programmatic actions that can be taken to reduce the level of carbon and other noxious substances that are continuously spewed into the environment. The remedies are known. The world seems very ambivalent when it comes to ideas about changing society's economic arrangements and cleaning up the environment even though the consequences of not acting are dire. This ambivalence can only be assuaged by educating the public on the positive benefits of transitioning the world-system to a new form. It is in that context that society will see the rationale for moving to a new economic configuration.

The forgoing issues lay bare the challenges faced by modern society today as it attempts to evolve and transition to a new social contract, a contract that focuses on preserving the environment and producing economic equality among the population internally and globally. Time is of the essence in addressing this issue. We must see the hurtfulness of the carbon-based economy and move to something that is more compatible to sustaining life and the planet. What that something should look like is the subject to be addressed in Part 2 of this book.

Bibliography

Board on Atmospheric Sciences and Climate, N. R. 1983. *Changing Climate: Report of the Carbon Dioxide Assessment Council.* Washington: National Academy Press.
Eisenstein, C. 2007. *The Ascent of Humanity: Civilization and the Human Sense of Self [Kobo version].* Kobo.com.
Flannery, T. 2005. *The Weather Makers: How We Are Changing The Climate and What It Means for Life on Earth.* Toronto: Harper Collins.
Intergovernmental Panel on Climate Change. 2014. *Climate Change 2014 Synthesis Report.* Geneva, Switzerland: Intergovernmental Panel on Climate Change.
Nordhaus, W. 2017. *Integrated Assessment Models of Climate Change.* Cambridge MA: The National Bureau of Economic Research.
Program, U. S. 2017. *U.S. Global Change Research Program Climate Science Special Report Vol. 2, Summary Report.* Washington DC: USA Global Change Report Program.
Rich, N. 2018, August 1. Losing Earth: The Decade We Almost Stopped Climate Change. *New York Times Magazine.*
United States Global Change Report Program. 2017. *Climate Science Special Report: Fourth National Climate Assessment, Volume I.* Washington DC: Global Change Research Program.
Wallerstein, I. 2004. *World-Systems Analysis: An Introduction [Kobo version].* Kobo.com.

3
TECHNOLOGY
The Trojan Horse

The advancement and rapid growth of technology throughout all spheres of life is a major factor in the accelerating need for wide-ranging and deep change to the social architecture and world-system that supports it. The present social and economic system is no longer able to accommodate the new technologies as they rapidly become established and incorporated into everyday life. There is no doubt that technology is advancing rapidly and becoming more invasive in human relations than ever before. The technological revolution, particularly digital technology, is now squarely embedded in the structure of individual and social life, and it will be even more invasive in the near and long-term than it is today. As Dyer-Witheford et al. tell us when speaking about the future of technology, specifically artificial intelligence, ' . . . it could also intensify or expand in a transition either to a significantly transformed capitalism, or to a radically different social formation' (Dyer-Witheford, Mikkola Kjose, & Steinhoff, 2019, p. 11).

The aim of this chapter is to provide a glimpse of the scientific and technological advances which the next great social transition will need to accommodate. Further, it will discuss the challenges the new technologies present to social and political life. And while the future is undoubtedly going to see huge increases in both quantity and quality of technology, let there be no mistake, society is past the point of no return on the question, so we need to embed them in a social architecture designed to our specifications and not left unregulated.

Massive technological change has initiated a new form of globalization. It is not just the outburst of population numbers or the colonization and exploitation of a county's natural resources, although some of that may still be going on today, but the digital technological revolution has sped up communications significantly, allowing for speedier transactions around the world. We are now truly a global village. And while it can be argued that the new technologies provide a positive force in daily life, they are also causing great disruption through the replacement of humans in the production process and significantly affecting other domains of social life. This turbulence is occurring not only in the manufacturing sector but also in the service and administrative industries. It must be said that the magnitude of these forces is unique to today's society. Although earlier civilizations faced disruption through the introduction of new technologies to their way of life, they were not as antagonistic to the social system as they seem to be today. Earlier technological advancement demanded more human labour where today they are labour saving and hence disruptive to the economic system. The advent of new technologies in earlier times produced more jobs, not fewer, but that is not likely to be the case any longer. Humans in the new age will need to grapple with the whole idea of life given that work and job have been the primary venue for gaining life satisfaction in the past.

Human development has been driven by prolific technological advancement. Most recently, we have witnessed the computer and digital revolution even though its early beginnings date back to World War II. It is an understatement to suggest that advanced technology has changed our individual and collective lives dramatically, just think about the smartphone; it has had a profound impact on how we live everyday life, and it is likely to become even more invasive as time goes on. Technological innovations, both mechanical and digital, have had great consequence to the human condition. The combustion engine has had a major impact on climate change and global warming. Advances in new technology will be a driver of the next great sociocultural revolution.

What has led us to this point in history? Technology, particularly digital technology, has reached a new threshold that is putting overwhelming pressure on the present social system. The new technologies are so life changing that they are a prime factor in the social anomie that is bursting out into demonstrable social chaos. Technology is instrumental in reducing employment in manufacturing. The new technologies are not labour intensive but rely heavily on highly skilled people unlike manufacturing which mainly employed moderately skilled and low-skilled labour. The social change achieved by digital technology is tantamount to the difference that steam made to the production process initially or the wheel provided to transport. These inventions dramatically changed the way people lived. Digital technology has had the same impact and it has only begun. Robotics has reduced human labour in the production process and the advent of artificial intelligence (AI) will replace human labour completely in most, if not all, economic sectors. Just think about what impact driverless vehicles will have on jobs in a number of sectors once that technology is perfected, and it appears to be on the immediate horizon.

The new digital technology jobs demand great mental dexterity rather than the manual skills required in industrial manufacturing. Even the manufacturer of goods such as autos that originally demanded a huge labour force now employ fewer workers and produce more cars than in previous times due to the increased use of robotics in the production process. Technological advance in the manufacture of goods and services has decimated working-class jobs, creating a permanent underclass. The future advancement in AI and artificial general intelligence (AGI), where robots learn from doing and then adjust their activities accordingly much as humans do, will spell the end of work not only in manufacturing and the service sector but also in administrative and other professions. The middle class workforce is now under siege and soon to shrink dramatically. The decentring of work will require a new system for distributing resources and necessitate creating new avenues for securing meaning in life.

The transition from manufacturing to the digital technological economy is creating great social dislocation in affluent societies. Plainly speaking, the new workplace requires fewer people in the production process and this fact is becoming all too clear to those who have been side-lined already by the new technologies. Addressing the social breakdown we are experiencing today will require new and different methods for wealth distribution for many individuals who find themselves left behind in the new economy. The present rhetoric of Making America Great Again by attempting to bring back jobs that no longer exist will simply exacerbate the present situation further and prevent society from getting on with the real job at hand, the transformation of the industrial society of the 19th and 20th centuries to something much different than exists now. Increasing social cohesion among the population will not be well served by the 'back to the future' approach to future social or economic development.

Industrial machine technology allowed humans to ravage the earth's resources at an accelerated rate, creating unprecedented environmental damage, including climate change and global warming. Technological advances are important to the evolution of humanity and continue to be important to this day. It will not be stopped, nor do I think it should be. There is great benefit to be obtained from the new technologies. We should not resist but harness them in a positive fashion and adjust our way of life to take advantage of them. The question remains, will humans adapt their sociocultural system to accommodate the advancing digital technology or try to resist its pervasive push. It would appear from the issues raised in this chapter that we are having some trouble in moving forward in harnessing the new digital technologies and the potential of AI to our benefit. The two concerns noted here may provide some clues to unravelling the forces humanity is now encountering and resisting.

It is suggested by many that we are entering the age of AI. Technological advance is now logarithmic and no longer slow paced. AI will soon gain complete control of the production process, eliminating the need for humans in the manufacturing of material goods. AI will also be felt in the services and administrative sectors as well as to the manufacturing of goods. Already our present society is on the brink of such advances as 3D printing and self-driving vehicles. People are now choosing to be implanted with microchips on which personal data are stored for a variety of personal purposes. It is also speculated that AI will not only be able to reproduce itself but also increase its capabilities and capacities over each new generation through experiential learning. This phase of AI is termed Artificial General Intelligence (AGI). It is expected to become self-reflective and learn experientially.

In other words, it will engage in a form of evolution without requiring the mechanism of copying and mutating genes, as do biological forms. This function will be generated through other means. AI will learn experientially and will continue to adjust and advance through trial and error. This is only the beginning of a great new age or Armageddon depending on your worldview.

Until now, humans have been directly engaged in, and in control of, the production process. We have invented and created technologies to solve problems of living through the production of basic goods and services required for survival, ensuring the proliferation and advancement of the human species. We have constructed a social and economic system that is centred on the job for advancing and organizing daily life. Work and job are the primary focus for social relations and, some would argue, they give substance and meaning to psychological life. When we reach the stage of giving over complete control of the production process to AI and robotics, what role and function will humans play in their day-to-day existence? What will replace work as the major mechanism for gaining psychological satisfaction? It seems clear to me that a whole new way of existing, and the construction of a new organizing system to support that existence, will be required. Paid work will need to be replaced with meaningful activity that is generated from within rather than directed from outside (e.g., the corporation) the individual. Early in modern human history, it was the church that directed and controlled day-to-day life and later it was the corporation. In the new technological economy, this function will become internalized to the individual. This transition should be viewed as the next stage in the emancipation of humans and not something to be feared or resisted. A call for the transition in individual and social life without a paid job will be the biggest leap made by humans to date, even bigger than transitioning from hunting and gathering to agriculture, and there is no guarantee that humans will be up to the task.

How we think and categorize things and events in society will need to be rethought. For example, our present patent laws may need to be reconsidered if society is to reap the full benefit of ideas in the transformation of the economy. What should remain in the private domain and what is beneficial to the public good and needs to be left in the public sphere, needs to be re-evaluated as society enters a hyper-technological existence.

We are, at present, biological creatures dependent on technology for our existence and further advancement as a species. Digital technology and AI will define human existence in the future, if it hasn't already. The coming scientific and technological advances that are under development now or being contemplated will provide humanity with instruments and tools for the continued progress of society if pursued and used wisely. But they will also provide the social and environmental context in which the next social transition will take place. Human existence can be thought to encompass three mega stages: biological, where life is completely controlled by biology; cultural, where culture becomes equal to or more controlling than the human biological evolutionary state; and finally, technological, where human life depends on technology for continued development. This later stage may eventually take the form of cyborg where technology becomes the predominant feature of intellectual progress. These technological advances may be both exoskeletal and endoskeletal. Although humans are not likely to be born with humanly created technological organs, they may be implanted with computer chips and other such devices at birth that will make them superhumans with many more capabilities than pure biological life can provide. Genes will also be manipulated before birth to create desirable features in humans. The social transition to which we are heading will drastically change not only the mechanical world but also the biological world we inhabit today. Technology and biology will merge to extend the human capability unimaginably. How have we arrived at this state?

Technological innovation has defined human existence since the dawn of Homo sapiens. Since leaving the trees for the savannah and becoming semi-bipedal beings, humanity has travelled through the stone, bronze, and iron ages, leading to agriculture and on to other forms of technological existence and culminating in the modern industrial era. Technology has been a major distinction between us and other forms of Homo since the dawn of hominids and it will continue to be a major contributor to human development long into the future. As Srnicek and Williams tell us,

> [I]f we are to expand our capacities to act, the development of technology must play a central role. As has always been the case, technology is the source of our options and options are the basis of a future that keeps us above the level of pawn.
> *(Srnicek & Williams, 2015, pp. 103, 104)*

To date, human technology has been mainly oriented to increasing the food supply to feed an ever-growing population and to create general health and wealth among the population. Human technology has focused on assisting people escape the frailties and vagaries of life. Society has recently entered the information age where new technologies are harnessed to advance life beyond food and wealth production in a myriad of ways. Transportation, energy production, communications, and information processing have led the way to the present-day technological advancements that have greatly benefited society to date. While these past inventions have certainly led to increases in food production and wealth creation, they have also contributed greatly to the general capability and progress of the human condition. Today's new technology is providing capacities to the individual and collective human condition that enhance the capabilities of each person and society generally. The simple addition of the personal computer has increased the capacity of each of us inconceivably in comparison to the technology available to our grandparents' and great-grandparents' generation.

To this time in history, technological advancement has been largely arithmetic. Today, technological innovation is logarithmic and only the speed of light, it would seem, provides constraint to its continued development. We are now using technology to create new technology. Recently it has been reported that AI has devoured all of humanity's antibiotic data and from analysis of that information has created a super-antibiotic. This is the first development of its kind and will lead to many more such advances.

There has been and will continue to be some potential dangers of putting so much emphasis on technology to advance the human condition. In fact, there are some present issues that require ethical consideration for the technology that is in use today. Social media platforms that threaten privacy and used inappropriately for nefarious purposes are a sample of those concerns. To date, society has experienced both positive and negative results from the adoption of digital technology. Certainly, the personal computer has added great benefit to the human experience. Perhaps one of the greatest benefits has been in the field of communications. Among many other functions, the internet allows us to send messages around the world instantaneously from our personal computer and, its extension, the smartphone. With such programs as Skype or Zoom, among others, we can communicate instantly face to face with anyone anywhere in the world. Much of our news and understanding of what is going on in the world today comes from social media. Using such platforms as Facebook, we can present ourselves and our opinions and beliefs across the globe. For better or worse, Twitter has provided a whole new arena for politicians to communicate their message to their constituents. Everything you say on Twitter becomes part of the permanent public record, much to the dismay of many who have used it inappropriately. But of course, there are always downsides to such inventions. Corporate espionage and stealing proprietary data are rampant throughout cyberspace.

Spying by all governments around the world is prevalent in international relations. That activity now includes spying through cyberspace. Cyberspace is also used by governments as a weapon against other sovereign governments. We are now, and have been for many years, involved in a cyberweapons race not only with the Russians, which has received the most attention, but also with China, Iran, North Korea, and perhaps others as well. Cyberspace is now where government intrusion into the businesses of competing jurisdictions is now carried out. We have witnessed the meddling by Russia in the 2016 US election, and it now seems difficult, if not impossible, to deter. Meddling of this type is also reported to have occurred in Europe and elsewhere around the world. It is like the proverbial iceberg; you only see the top third of it above the waterline. This kind of intrusion has caused some to argue that conflict has transitioned from ground, sea, and air combat to cyberspace warfare. Sanger (2018), in his book *The Perfect Weapon*, contends that the United States and Israel shut down Iran's nuclear program through cyberspace warfare. Governments routinely hack into the computer systems and utility installations of other governments. Hacking goes as far as placing malware in the computer networks of sovereign country utilities and other systems, which can be activated in the future to disable those networks and cripple the foe when required. It is hypothesized that there is an ongoing attempt by Russia, and perhaps other countries as well, to destabilize liberal democracies by invasion through cyberspace. David Sanger identifies several areas of political activity and social discourse that are being intruded upon through cyberspace warfare. There have been many attempts to hack utility infrastructure, such as the electrical grid, nuclear power plants, and water distribution systems in many countries, thankfully without disastrous results so far. The

Pentagon and other governmental institutions have also been targeted for attack, once again, without huge negative effects, as far as we know. Fake news that is directed at fostering disruption among the population is being perpetrated through many social media platforms on the internet. Voting registration records are being attacked by reprehensible agents through cyberspace with the aim of eliminating voting by certain populations who are likely to vote for a specific type of candidate. Political party data bases are stolen and transported to opposing political groups with the purpose of providing compromising information on opposing candidates. It has been made clear that many emails were stolen from Hillary Clinton's email server during the 2016 election campaign and made public to embarrass her and to change voting patterns. There is also suspicion that voting stations were attacked during elections with the hopes of influencing the results of the election. Fortunately, this does not appear to have been achieved, but unless stronger defences are created, this sabotage has the potential to ultimately change election results and severely damage democracy. Some of these attempts at nefarious actions have been more successful than others but these practices will continue unless coordinated steps are taken to ensure they are thwarted. It needs to be pointed out that the United States has created government departments that use cyber technology to attack other countries as well. The United States has not only created such capacity but also using it for some time. All this has been going on nefariously without governments revealing their activity. If this activity is to have deterrent value, as has been the case so far with nuclear weapons, then all participating governments will need to divulge their activity and begin to negotiate the rules of cyber engagement. Governments around the world need to disclose their nefarious practices before the goal of achieving an armistice can be negotiated. To date, no one will confess to these activities, but some are starting to hint at their existence.

Society needs to focus its attention on rectifying this type of negative use of cyberspace and the technologies that perpetuate it. Many present and former members of the intelligence community see cyberspace conflict as the next platform where warfare between and among nations will be conducted. Infrastructure and commerce can be virtually shut down through placing malware implants in the computers and systems of targeted countries. Despite such negative experiences, technology in cyberspace will continue to develop and become ever more invasive in human existence, much like the previous arms race between and among countries. It is projected that the cyber warfare race will continue for some time yet, so it is becoming increasingly necessary for the world to develop the basic ground rules for its use and, hopefully, eliminate the present and potential nefarious uses made of it.

Despite the difficulties and challenges referred to here, there are some very interesting advances being made in the technological world that will dramatically change how we live life in the future. Already, there are numerous implants that can aid the deaf, blind, and physically challenged live richer lives. People who have been paralyzed are now walking upright due to robotics. The promise of this kind of life-enhancing technology is unlimited.

Toward the completion of this book, much focus in the technological world has been placed on the creation of driverless vehicles. This development may seem like a small advancement given some of the earlier major achievements in the technological world, including human flight, but driverless vehicles will have considerable influence in how we live everyday life in the future. When one looks at it closely, one begins to understand the significance of such an achievement to human existence; hopefully, there will be no more auto accidents, which claim a significant number of lives on the roads each year. The efficiency of moving goods and services continuously given that the operational system does not have to rest would be a notable achievement. The downside is immediately apparent, however. The loss of jobs will present post-modern society with a sizable obstacle to overcome. Think of all those who make their living driving vehicles of one description or another. Worker displacement will be severe with the widespread introduction of this new technology, and we are just around the corner for that possibility to become reality. There are several manufacturers who have prototypes in operation today. Economic displacement is a huge outcome of this new technology as the automobile was to the horse-drawn carriage in the late 19th and early 20th century. The transition to the AI age will require us to rethink the role of work in the lives of people and the present method for distributing wealth. More will be said about that subject later.

The recent development of drones cannot be easily overlooked for their potential impact on daily life. In conjunction with driverless autos, drones have the potential to give humanity its first personal flying machine that will be able to operate much like their present vehicle but in the sky rather than being permanently attached to the ground. While this will provide one answer to the congestion and gridlock problem we now face on our major roads, it will transfer that problem to the sky above so that issue will need to be addressed and resolved before this potential becomes reality. But make no mistake, it will be solved in short order and a whole new mode of transport will open to society. Drones can be used in a variety of ways including delivery of packages. They are already being used for geological mapping and, unfortunately, by the military for precision unmanned strikes.

A most interesting recent leap in the technological world is 3D printing. 3D printing is computer-based three-dimensional computer-controlled printing using a variety of materials in the construction of an object. Products of any shape or size can be constructed from a digital model. Although in use today, 3D printing is still in the experimental stage and not useful on a widespread commercial basis yet. To date this new technique has not had a major impact on the production process that makes the goods we enjoy on a day-to-day basis, but if this method meets expectations, it will eventually revolutionize the production of all types of goods. What is significant to this discussion about the potential of 3D printing is the effect it will have on the manufacturing and retail industries. I suggest this method of manufacturing will dramatically reduce employment in both manufacturing and retail sectors. The downside of 3D printing has been the nefarious uses it has been put to so far. Most recently in the United States, plans for printing handguns has been uploaded onto the web by a self-described anarchist entrepreneur. The problem with these guns, in addition to the lack of need for more guns in US society, is their anonymity. Plastic guns created through 3D printing will be undetectable and untraceable using present security technology. These guns will also be available to people who are unable to purchase regular firearms because of failed background checks. Fortunately, this practice has been stopped by a court injunction until the issue can be thoroughly vetted and determined to be made allowable or not under the US Constitution. Hopefully, it will be permanently stalled. The major problem this example points to is the lack of response to this venture by the US government. The injunction to halt this practice was lodged by non-governmental and state officials but not the federal government. It demonstrates clearly that legislation is not keeping pace with the introduction of new technology. Governments across the globe will need to get in front of the curve of advancing technology, or the consequences will be more than just concerning – they will be calamitous.

Another more recent invention of significance is the creation of meat in the laboratory, thereby moving human existence beyond the direct relationship between meat for protein and animals. This will move humans another degree away from their biological existence much as AI is doing in other spheres of human existence. Humans have spent much time and effort in cultivating animals and plants for food. Since the early days of modern humans, domestication of wild plants and animals has been a major preoccupation. The domestication of plants and animals allowed humans to leave their nomadic way of life for a more sedentary lifestyle, thereby permitting a significant increase in their numbers. This represents the first great leap forward in the process of technologizing the human condition. That is not to suggest that fire, stone tools, and the discovery of metal were not huge technological innovations themselves. Early on, humans possessed primitive tools which distinguished them from other hominids, but the introduction of sedentary agriculture set humanity on a sophisticated technological path unlike their hominid relatives.

Until now humans have flourished because of their ability to grow enough food to feed a large population across the world. Why we see hunger in some parts of the world today is not because of our inability to grow enough food but a result of distribution issues created by humans themselves through the present economic system. If the creation of laboratory meat becomes prolific, think of the uses the land that is now devoted to agriculture could be put, perhaps in a manner that is more environmentally sustainable. This change, like driverless vehicles, would be a game changer in the structure of human existence. I suspect that human labour input in food production and distribution would also be minimal; at least it would occupy many fewer people than our present food production and distribution system involves. Technological advance has, until now, created more jobs than it has destroyed but that may no longer be the case. It is suggested that one robot replaces six workers in the production process.

As if the technologies briefly outlined earlier are not enough to boggle your mind, there are still areas to explore that have the potential to alter life dramatically. The next great science and technological frontier that has the potential to drastically affect human life for better or worse is quantum gravity and dark matter. Digital technological and AI advancement will take another great leap forward once the mysteries of quantum gravity and the dimensions and components of dark matter are unlocked. Energy production and other significant domains (see Tegmark, 2017) are likely to become unlimited in their benefit to humankind once these scientific wonders are solved. Although the scientific community has witnessed the results of quantum mechanics, no one truly understands its components and workings. Dark matter is also such an enigma. What is clear is that once understood, science and technology will have unlocked another area for advancement. It may be as important or perhaps even more significant than the creation of AI.

Not to be overlooked are the great advances made in genetic research. The mapping of the human genome was completed in 2003. This accomplishment has opened the world of genetics to a variety of purposes. Humans are about to have the capacity to direct their biology artificially, where, until now, change has been limited by random gene mutation and natural selection. It is now possible to purposely direct our biology in ways we have previously not dreamt (Metzl, 2019). This discovery presents several profound ethical issues that will need to be addressed before we get much further down this road. The ethical dilemma this discovery generates will make the discussions that now rage on the right to life and the abortion issue seem quite modest and uncomplicated by comparison.

Since humans domesticated the wolf, we have selectively bred 340 varieties of dog. This has been accomplished by selecting attributes for breeding and then combining those to produce a new strain of K9. Through gene mutation and the domesticated version of natural selection, we have been able to accomplish this feat within constraints. Now that we have a complete map of the human genome, natural selection can be circumvented in favour of direct manipulation of genes. Anything is now possible, including extending life itself. This development can lead to scary but potentially wonderful outcomes. Just how far we will be able to extend life is now an open question.

Although I have briefly mentioned the issue of work reduction due to technological innovation in the workplace earlier in this chapter, this subject deserves a more detailed examination of its significance to social and psychological life. The new technological revolution which includes the potential for the creation of AI may continue to dramatically affect human labour in all sectors of the economy. The centrality of work in society may be on the way to redundancy given inventions such as driverless vehicles and AI. The AI advances that appear to be on the horizon has led Tegmark to muse,

> (e)verything we love about civilization is the product of human intelligence, so if we can amplify it with artificial intelligence, we obviously have the potential to make life even better. Even modest progress in AI might translate into major improvements in science, disease, injustice, war, drudgery and poverty.
> *(Tegmark, 2017, p. 110)*

Technology is such a part of our everyday existence that some scholars and pundits suggest that the continued development of AI will produce a new species of human, resulting in an individual that is as much technological as it is biological. If this were to come true, and given our technological accomplishments throughout history there is no reason to doubt it (much of our technology today was science fiction turned into reality); humans are on the precipice of a great transforming journey driven by science and technology mainly through the exciting new field of AI. To give but one example of what is new in the field of AI, an excerpt taken from a larger article describes what Google's London-based DeepMind research project is working toward. The title of the article is, *Google's DeepMind Made an AI That Can Imagine the Future*.

> Imagining the consequences of your actions before you take them is a powerful tool of human cognition. When placing a glass on the edge of a table, for example, we will likely pause to consider how stable it is and whether it might fall. On the basis of that imagined consequence we might readjust the glass to prevent it from falling and breaking. This form of deliberative reasoning is essentially 'imagination', it is a distinctly human ability and is a crucial tool in our everyday lives.
>
> If our algorithms are to develop equally sophisticated behaviours, they too must have the capability to 'imagine' and reason about the future. Beyond that they must be able to construct a plan using this knowledge. We have seen some tremendous results in this area – particularly in programs like AlphaGo, which use an 'internal model' to analyze how actions lead to future outcomes in order to reason and plan. These internal models work so well because environments like Go are 'perfect' – they have clearly defined rules which allow outcomes to be predicted very accurately in almost every circumstance.
>
> *(Tauber, 2017, p. 1)*

Reasoning and planning are two of the distinctive qualities that define humanness. We may be no longer alone in those competencies. When machines become fully capable of learning from their experiences, they will become more autonomous and control their own (and maybe our) destiny.

On the horizon is the creation of Artificial General Intelligence (AGI) as the aforementioned quote suggests. The major factor here is the ability of AGI to learn and advance its own condition and competencies. In addition to, or as part of, the ability to learn, AGI will incorporate the idea of intent and the desire to fulfil that intent. There are fully functioning machines today that can complete tasks, as well as or better than humans. Most of these machines are single focused, that is, they perform a solitary task. The future of AI portends the creation of general robots that perform unlimited tasks that are guided by the intent and the desire to seek out the next task that leads to the accomplishment of an overall predetermined goal. When AI can function in this manner, it will have reached a state of AGI and will have acquired this unique human feature. The ability to learn experientially with the intent of changing the future is what sets present-day humans apart from the previous stages of hominid, and this is the path AGI research is pursuing. Perhaps the most intriguing aspect of this new development is achieving the function of human cognition. If this were to be achieved:

> AI would become a cognitive analogue to the means of transport and communication – *the means of cognition* (italics the authors). The means of cognition would be a new layer of technological infrastructure interlaced with both the means of production and the means of transport and communication. While capitalist production has always relied on the human capacity for cognition in both conception and execution, an infrastructure of AI would allow the distribution of cognitive and perceptive tasks to machines, which would perform them in different, machinic ways, with potentially revolutionary effects on the mode of production. . . . AI is not only significant for its potential to automate work but also in so far as it could become part of the general conditions of production. If the means of cognition are established, production will certainly become increasingly automated, and it will become so within an environment where intelligent machines are perceiving, learning and communicating.
>
> *(Dyer-Witheford, Mikkola Kjose, & Steinhoff, 2019, p. 88)*

There are some among us who would not want society to venture forward into the creation of such technologies, and there are some very good reasons for their foreboding. Creating an intelligence, whether artificial or not, presents a change to human society that is unequalled since the beginning of the species Homo. Such previous transitions as agriculture may have been a bit daunting to hunters and gatherers, but that accomplishment does not compare to the magnitude of the creation of Artificial General Intelligence. This new technology has the potential to completely alter the biological, psychological, and social functioning of the human being. There are some serious consequences of this new technology that did not accompany the advent of sedentary agriculture, although many of the diseases that were created during the arrival of agriculture should not be dismissed. That said, I don't believe there is any turning back at this point. AGI that rivals the capacity of humans, and maybe beyond, is inevitable. The human species is dominant in the world today simply because of its inventiveness and, perhaps most importantly, its curiosity. To think we could turn back at this stage in the development of AI is unthinkable. It is not only inevitable but also desirable that technological development will continue to occupy a greater place in human life in the future. The question is not 'will it become increasingly important to human life' but 'how can it be incorporated into everyday life and become a benefit to all of us'? While we are singularly focused on creating such technology, we are slow to develop appropriate regulation to control it or set goals for its creation.

The fact that machines may gain unimaginable competence produces a crisis for society in that there is immediate need for ethical and goal determination prior to the creation and proliferation of these pending technological achievements. Will technological development end up leading and controlling society, or will humans always be in control of technological innovation? This is a major question to be addressed and one that could determine the fate of humanity.

I am not suggesting a sci-fi Armageddon is at hand, as it is often portrayed in science-fiction literature and the cinema, but I do conclude that humans and technology will become symbiotic and more integrated in the future. As Max Tegmark suggests, '(w)e can think of life as a self-replicating information-processing system whose information (soft-ware) determines both its behavior and the blue-prints for its hardware' (Tegmark, 2017, p. 35). Even today most people (particularly the young) are clearly attached psychologically, if not physically, to their personal communication apparatus. The World Health Organization (WHO) has recently created a new mental illness designation for those who are 'addicted' to their electronic equipment. Perhaps in the future, this relationship will not be viewed as a mental illness but simply an extension of who we are. This relationship will likely occur gradually and be welcomed by the human community and not viewed as some Faustian bargain ending in the collapse of all life on earth. If there is to be a collapse of life on earth, it is likely to occur because of our inability to engage technology and deal effectively with many of the social and environmental problems, such as global warming, to which we are now confronted and seemingly unable to address effectively. To what extent and in what form this proposed symbiotic relationship with the new technologies will take is not totally clear at this time. Some future gazers are hypothesizing that we may witness a complete metamorphosis of humans into cyborgs, that is, humans implanted with an array of technologies from birth, which, in effect, will create a whole new species of humans. We are not talking only of artificial technology such as silicon chips but also of gene manipulation when we speculate about technically altering human life. We implant artificial remedial apparatus such as knee, hip, and shoulder joints into our bodies on a regular basis without hesitation. The next step in artificiality is the implantation of developmental equipment to enhance human performance, allowing humans to reach beyond their natural biological limitations. Once that divide has been crossed, some humans may decide to receive such a transformation and some not. Humanity may divide into two divisions. This is not to suggest that humans not imbued with technology will live a substandard life in subjugation but simply a different life in many ways from those who embrace a technological fusion. In essence, there is likely to develop two separate human paths. Homo has experienced many branches in the past and this may become one more. I'm not suggesting anything nefarious here but simply that over time people will come to see the advantages of the new technologies and join in the conversion even if it takes generations. We must remember that there were other forms of Homo, such as Homo erectus, Homo heidelbergensis, and Neanderthal to name the most prominent, who lived on earth many years ago. We know that Neanderthal lived simultaneously with humans before becoming extinct. Some researchers have claimed to have found some Neanderthal genes in present-day humans, but this is still speculative, not confirmed. Notwithstanding that controversy, those other branches of Homo died out or merged, leaving Homo sapiens the lone survivor on the planet today. The logical conclusion of this trajectory they conclude is a new human distinct from Homo sapiens much like we are distinct from Neanderthal. Perhaps it will be dubbed 'Homo techno'. Admittedly, the scenario presented here is likely to be some time in the distant future. That said, it is difficult to put a time frame on it because technological advance seems to be speeding up considerably. If I had to guess how long it will be before the possibility of the aforesaid scenario taking shape, I would suggest sometime before the end of the 21st century. Predicting the future is tricky business and not likely to be accurate, but the scenario presented here has some basis in fact given what the technologically focused world is engaged in at present. There are some life-changing technological advances that are on the immediate horizon that have considerable probability of becoming reality soon. The extension of these advances could very well take us to the cyborg phase of human existence. The most critical element in the decision to go this route or not is that it be clearly thought through and cognitively understood and agreed to by society. This is not something that should be left to the commercial sector; their incentives for making an affirmative decision would not be the right ones. The individual decision to travel down this path or not must be accompanied by strong policies and laws that negate the potential for any form of discrimination to develop in society based on the decision to go one way or the other. If this scenario scares you, that's good; it represents a huge departure from our human story. That's what makes this discussion so troublesome. Much would need to change in our present social structure before we are in a position to even contemplate such a technology-infused human life, but we need to start thinking about it now given the speed with which our technological universe is now travelling.

There are also much darker possibilities than Tegmark's optimistic view. AI also has the potential to be a life-destructing force. Many of these possibilities are presented in Hollywood films and science-fiction novels. The possibility of robots and cyborgs being created and used for evil purposes is possible but highly unlikely. Certainly, it is unlikely if present humans think futuristically and begin now to discuss and determine on what ethical basis this new technological world will unfold. Much of the work in AI is collaborative across universities from around the world and not being undertaken in some underground hideaway. Conferences are frequently convened where not only technological advances are being shared but also the ethics of AI is under constant discussion. Unfortunately, governments appear to be very slow to follow up with regulations or even recognize the inherent dangers presented by advances in technology. In fact, China and the United States are reducing regulations on the AI industries as each country races to dominate the field (Dyer-Witheford et al., 2019). They seem to be more concerned with the commercialization of these new devices than they are with the ethical implications they carry with them. It is only in hindsight where we begin to see the morass created by some of the present technology, like Facebook and other forms of social media. Ethical consideration followed by regulation must precede implementation and not leave society with the task of catching up. This is particularly the case when we begin to consider technological implants and gene manipulation.

Beyond the potential sinister uses of AI, there are much deeper problems to be overcome to ensure AGI is a blessing and not a curse when it finally arrives. Many of these questions come in the form of social and psychological adjustment by humans. Issues such as how humans find meaning in life without work becomes an urgent question to be addressed. Work has been the dominant, if not the exclusive, life-organizing mechanism for humanity since the beginning of the species. What are the root causes of anomie and lack of social cohesion in society today? One of those root causes lies in the replacement of meaningful work by technology and automation of the workplace. Loss of purpose in life through the automation of work is an unintended consequence of the technological world that is replacing humans in the manufacturing process and creating few meaningful jobs at one end of the spectrum but leaving only meaningless jobs for the majority of the working population at the other end. As Viktor Frankl tells us:

> Every age has its own collective neurosis, and every age needs its own psychotherapy to cope with it. The existential vacuum which is the mass neurosis of the present time can be described as a private and personal form of nihilism; for nihilism can be defined as the contention that being has no meaning.
>
> *(Frankl, 1992, pp. 113–114)*

Populism promises to reverse course and return to a period in history where everyone had a purpose in life no matter how unbearable it might have been. Men laboured at the office and in manufacturing plants and were considered the family's breadwinner, while women took care of hearth and home. This arrangement was very neat and tidy but not always fulfilling. Logotherapy directs its attention to the 'essential vacuum' resulting from the conditions described herein. As its originator, Viktor Frankl explains:

> The existential vacuum manifests itself mainly in a state of boredom. Now we can understand Schopenhauer when he said that mankind was apparently doomed to vacillate eternally between the two extremes of distress and boredom. In actual fact, boredom is now causing, and certainly bringing to psychiatrists, more problems to solve than distress. And these problems are growing increasingly crucial, for progressive automation will probably lead to an enormous increase in leisure hours available to the average worker. The pity of it is that many of these will not know what to do with all their newly acquired free time.
>
> *(Frankl, 1992, p. 133)*

It may be that distress is as rampant today as boredom was in its time. We can see distress in society's adventure into populism and nationalism. The aforementioned quote raises important issues humankind needs to resolve if it is to get past this foray into populism and nationalism and its futile attempt to return to some idyllic past. As Frankl clarifies, '(s)uch widespread phenomena as depression, aggression and addiction are not understandable unless we recognize the existential vacuum underlying them' (Frankl, 1992, p. 114).

Continued automation of work will create two problems for individuals in modern society. The first is answering the question, where will I obtain an income or sufficient resources to live a comfortable life, and second, how will I find meaning in life without paid market-based work? Work in the technological age, as much as we may want to think otherwise, will diminish in importance for gaining sustenance and, perhaps more importantly, for providing meaning to one's life. We have been on this diminishing path for some time. I have dealt substantially with both these subjects in other volumes (Reid, 1995, 2017), so I will not go into them extensively here. Suffice it to say, work makes up a huge part of every person's present-day life and in the organization of society generally. In the capitalist economy, and for other economic systems as well, work is what gives meaning to life. So, one cannot diminish its importance to the discussion. On the contrary, work has been so substantial it needs to be given treatment in a volume of its own. What can be stated here is that meaning in life will need to be found in new domains, mainly in leisure and other meaningful non-work activities. A life without paid work for some would need to be accompanied by a major change to the social system to accommodate it. This change to lifestyle would need to be part of the transition discussion held on a society-wide basis.

What system can be constructed to share the resources that each of us need for securing food, shelter, and other life-giving commodities? These are major issues and questions that need to be addressed before going much further into this brave new world. There is no doubt the fabric of society and the tasks around which it has organized itself until now would need to be discarded and replaced with a social structure that is appropriate for the AGI age where most work is turned over to AI and robotics. Tegmark identifies a number of factors that contribute to the wellbeing of humans that would need to be replaced in a workless world:

> [A] social network of friends and colleagues; a healthy and virtuous lifestyle; respect, self-esteem, self-efficacy and a pleasurable sense of 'flow' stemming from doing something one is good at; a sense of being needed and making a difference; a sense of meaning from being part of and serving something larger than oneself.
>
> *(Tegmark, 2017, p. 152)*

Although these values have traditionally been found through work, they can also be found in leisure and in meaningful non-work activities. Many of those not in the workforce at present, including retired people, find great meaning in their everyday lives. Research (Reid, 1995, pp. 129–148) reports that a preponderance of people find purpose in life outside of their job. The connection of these values solely with work may be one of the major myths created by the capitalist system for its own benefit and will need to be rethought. More will be said about this subject in Part 2 of this book.

In addition to the meaning of life question produced by automating the workplace, there are additional and more profound questions that the advancement of AGI will produce. What are the principles on which AGI is constructed? What are the goals of AGI and how are they implanted in new technologies? Finally, what is the ongoing mechanism for establishing and updating ethics that will direct the development of AGI?

In a sense, the ethical system that is demanded by the development of AGI will provide the framework for creating principles and goals for its use, so let's start there. The ethical system will need to address the question of goals and how they will be established. Our history with technology has been to address ethical questions only after we encounter problems through their use, not prior to their creation and implementation. In the case of AGI, questions of ethics need to be resolved before development begins in earnest; otherwise, technological advance is so fast moving that we run the risk of never catching up once the race has begun.

Given the implications of AGI that portends to completely change human existence in every way, the question of who should be involved in establishing the ethical framework is a critical one. Under normal situations I am usually in favour of encouraging as many people as possible to be involved in the decision making process and to have equal voice in that process. Given the complexities of this subject, however, there is great need for special expertise that understands the potential benefits and dangers to the globe and humanity of high technology, so special weight must be given to the science and philosophical communities in this discussion. At the very least they should be called upon to explain the nuances and ramifications of technology to the human system. Other physical and social science disciplines will need to have considerable input into the process given they have special expertise that will need to be taken into consideration when analyzing the consequences of any configuration of AGI. Additionally, these communities must communicate their deliberations widely to the public and other groups who have prodigious interest. The importance of this issue is so great that the process of ethical decision making must be methodical and completely transparent. Widespread public education to the intricacies of AGI and what the costs and benefits are likely to be to individuals, society, and to life itself must form part of the public discussion. In conjunction with this dialogue, the value of science to society and the decision making process must be reconstituted so that confidence of the general population and politicians in the value of special expertise and the science community is built up again. For the most part, expertise and science has been summarily dismissed by a large part of the population and their political leaders. This has been the case in the global warming debate, for example. The same scepticism exists when it comes to many of the large technical issues in society today. This attitude must be overcome as we move into the brave new world of advanced AGI and robotics. Decision making in this domain must once again rely on the special knowledge produced by expertise and science. Science cannot make decisions on these and other matters, but it must certainly inform them.

Once the education of the public regarding the ramifications of AGI to society is determined to be adequate, the discussion can then turn to the society-wide goals for AGI and the principles on which it will be implemented. The social organization system through which AGI will be executed will need to be addressed at great length. Such issues as how wealth and resources emanating from the use of AGI will be distributed must be clarified. The extent of gene manipulation allowable in unborn children and the type of implants available for endoskeleton use are examples of important issues which society must determine. Defining the final arbiters for each case is also an issue for debate.

The new society will need to grapple with problems much different than those of the industrial age. What tasks will preoccupy humans in the AGI age? The day-to-day activities for each of us in the AGI era will need to be supported by society through education and counselling. Leisure will become the focus of attention in the AGI age, much as the economic system and job-skill training was to the industrial epoch. When I speak of leisure here, I don't simply imply frivolous acts of time consumption as we often think of it today but meaningful activities that give substance to life. Stebbins (1982) calls it 'serious leisure'. He identifies such activities as volunteering, hobbies, and being an amateur as constituting the domain of serious leisure for many people. People today are very familiar with those concepts and much of their present life is taken up in those pursuits. Much public good is provided by people who volunteer their time in a variety of ways for the benefit of others. The question of gaining meaning in life is not whether one is employed in the market economy but how productive and useful we are to ourselves and our community.

Mair identifies civil leisure as an important sub-activity of serious leisure. She describes this type of meaningful activity as 'leisure based in resistance concerned with creating public discursive space. This definition is exemplified through both the Occupy Wall Street and Arab Spring movements' (Mair, 2002, p. 213). The ideas contained within the concept of civil leisure is exactly what is meant in the aforesaid discussion on public involvement in creating the framework for AGI ethics and in setting the principles and goals for its development and evaluation. Mair's intuition regarding civil leisure may also be a main platform for providing meaning to life over the long-term in a transformed society. Although the major activity in her definition of civil leisure is 'resistance', it may be broadened to include more contemplative and participatory actions and less confrontational forms of political and social involvement. It implies intense mass participation in determining the goals and social organization strategy of society and then its continued monitoring and adjustments. This type of activity will be of paramount importance in the new age, perhaps the major focus of civic engagement. Given the low numbers of voters turning out for elections throughout the liberal democratic world these days, the refocus on civic life would be a welcomed development in the western world.

The preceding short discussion including examples of the rapid development of digital technology over the recent decades, together with the current expansion into AI and aspirations for AGI, will give the reader a sense of where humanity is headed in the future in terms of its social and cultural evolutionary path. Some may argue that what we are witnessing today is not evolution in the biological sense, but human development based on the symbiosis of silicon technology with biological evolution through gene manipulation. What is certain in this relationship over the near and long-term future is the increased dominance of culture over biology, if it isn't already, leading to a profoundly changed human and social existence. The decision to pursue this direction has been taken, and we have already, knowingly or otherwise, embarked on a path of no return. We have thrown our lot in with the technological revolution and there is no turning back. What is left undetermined is whether humans will embrace and control it or let it control them.

The conversion to a deeper embrace of technological dependency and social alteration will not evolve smoothly but none of the preceding transitions have been without their upheavals, either. The introduction of a new life-altering technology in civilization has caused great disturbances historically. Inventions in the world of communications such as the printing press caused enormous political and religious upheavals lasting 130 years. The Bible and other sacred books and teachings, which gave meaning to life's existence, not only became available to everyone but went through many meaning changes depending on the point of view of the publisher, causing great divisions between and among populations. New religious tribes were born that pitted their version of life against others. We are in a similar position today with the advent of the internet and its many social media platforms and networks such as Facebook and Twitter. They too are subject to providing false interpretations known as fake news in today's parlance. As in the case of the printing press and the printing of the Bible in the 17th and 18th centuries, social media is aiding in creating separate tribes in society that are finding it increasingly difficult to share the same social space. Although there have always been differences of opinion in politics, there is now a significant chasm between the numerous tribes that have recently emerged. These separate groups find it difficult, if not impossible, to speak to one another. It appears difficult for them to share the same reality of society, which is creating political gridlock. The internet and social media are only examples of the massive influence of the technological world we live in. And, while the invention of the internet and the international communications networks provide positive influences, it must be recognized there is a dark side to this technology that may affect our lives negatively as well. Think of the misuse of social media in influencing, maybe even turning, elections through the explosion of 'fake news'.

Certainly, the direction new technology takes will require a great deal of deliberation regarding the developments themselves and the priority needs of society. A large part of technological expansion, as alluded to previously, will be to establish mechanisms to make those priority determinations. This is by no means an easy task but one that will require a sophisticated and an all-inclusive citizen participation strategy, one that values scientific fact and special expertise. But this is only part of the work to be done.

The foundation on which technology developed in the past was based on entrepreneurial spirit. If one person has a good idea, they are encouraged to move that idea forward to eventual adoption by society. With few exceptions, there was essentially no evaluation of the consequences to society of the invention. This process could be said to emulate natural selection in the biological world. Market-forces advocates argued that their success on the market demonstrated what was good for society. Whether this theory was ever valid can be argued but suffice it to say that this notion will not be adequate for creating technology in the future. Today, the eventual consequences of technologies to society demand that the public control what paths to follow, not individuals. It is not just markets that are being affected by AGI and other new technologies, but life itself. This demands a new foundation on which to create and operate the public enterprise.

The new technological age we are entering, perhaps well into, will require a completely new social structure to accommodate the consequences of this new innovative way of life. For example, if the new technologies eliminate work as we know it today, a new mechanism must be constructed to distribute life sustaining resources to the population. Although this may seem logical as you read it here, this concept confronts the very basis of the capitalist system as we know it today. The capitalist economy is solely focused on growing the economy through work and the job. The new technological age as envisioned here will require the leaders of society to think outside the box and not double down on trying to fit the new technological age into the present social organization architecture. The tendency of the human community is to attempt to accommodate new initiatives in the present structure, but that approach will simply exacerbate the problems with which we are presently grappling but not resolving. We already see signs of these frustrations as we witness the advent of populist movements across the western world. Citizens are looking for the 'strongman' to fix the problems that are a result of today's technologies and their consequences. But those at the top of society may have a different set of priorities than the rest of us, so priority must be placed on creating an inclusive participatory strategy to address the issues so that everyone's voice is heard and mass society benefits from the changes that are made to the system.

It is important to note that there are other issues in addition to the technological ones listed here that contribute to the present upheavals in society, such as global warming, poverty, and the problem of inequality which will be addressed in due course, but technology is a major contributor to the social upheaval that society is experiencing at present and will continue to experience well into the future. Today's social disturbances will continue to grow until society recognizes that we are living in an altogether different world, one that requires a drastically different social organizing system to accommodate the ever-expanding technological world. We must stop resisting but embrace it and organize it to our benefit. This means taking control by taking it out of the hands of the corporations and regulating it through the people's government. Society must get serious about regulation. This would mean dispelling the notion that government is part of the problem and not a positive force for good but only needs to get out of the way and let capital do its business. This position now constitutes the 'big lie' that needs to be exposed for what it is – a myth. It must be remembered that Silicon Valley happened because of the venture capital provided initially by government, and it was not until the corporate world could see the potential profit the new technologies could eventually produce, were they willing to take the risk of investing (Mazzucato, 2014).

The potential impact, both positive and negative, of the new and pending technologies is so great that it must be developed through a cooperative and non-market mechanism, not the commercial market. Yes, there will be technologies that can be turned over to the market for development and distribution but only those that are proven not to have potential negative impact and a detriment to life. We have done that in the past with nuclear power, so we can do it again. It is important for society to not reproduce the ills of the 19th and 20th centuries where our technologies imperilled the health of the planet and the people on it. The control and production of the new technologies will need to occur within a wide-ranging overhaul to the world-system, and in concert with the other domains in the system, that will also need revamping at the same time.

Bibliography

Dyer-Witheford, N., Mikkola Kjose, A., and Steinhoff, J. 2019. *Inhuman Power: Intelligence and the Future of Capitalism [Kobo version]*. Kobo.com.
Frankl, V. E. 1992. *Man's Search for Meaning [Kobo version]*. Kobo.com.
Mair, H. 2002. Civic Leisure: Exploring the Relationship between Leisure, Activism and Social Change. *Leisure/loisir: Journal of the Canadian Association for Leisure Studies* 213–237.
Mazzucato, M. 2014. *The Entrepreneurial State: Debunking Public vs. Private Sector Myths [Kobo version]*. Kobo.com.
Metzl, J. 2019. *Hacking Darwin: Genetic Engineering and the Future of Humanity [Kobo version]*. Kobo.com.
Reid, D. G. 1995. *Work and Leisure in the 21st. Century: From Production to Citizenship*. Toronto: Wall & Emerson.
Reid, D. G. 2017. *Social Policy and Planning for the 21st Century: In Search of the Next Great Social Transformation*. London UK: Routledge.
Sanger, D. E. 2018. *The Perfect Weapon: War, Sabotage, and Fear in the Cyber Age [Kobo version]*. Kobo.com.
Srnicek, N., and Williams, A. 2015. *Inventing the Future: Postcapitalism and a World Without Work [Kobo version]*. Kobo.com.
Stebbins, R. 1982. Serious Leisure: A Conceptual Statement. *Pacific Sociological Review* 251–272.
Tauber, A. 2017, July 24. Googles DeepMind made an AI that can imagine the future. *The Next Web*. http//www.msn.com/enca/news/techandscience
Tegmark, M. 2017. *Life 3.0: Being Human in the Age of Artificial Intelligence [Kobo version]*. Kobo.com.

4
POVERTY
The great unequalizer

The growing separation of the economic classes has been clearly documented (Piketty, 2014; Piketty & Saez, 2003; Wilkinson & Pickett, 2009). The middle class in most if not all, capitalist countries, is diminishing in size, not because they are joining the upper classes but as a result of their weakening economic condition. The exception to this rule can be found in some Asian countries, particularly China, which, strictly speaking, is not capitalist in the traditional sense even though it has embraced some market concepts. That said, the Chinese economic miracle, as it is often referred to, continues to contain large numbers of citizens mostly in the countryside that remain in poverty, even though the standard of living for many has improved, particularly for those who have migrated to the cities. As in the West, China is also producing some very wealthy people. In most of the other parts of the globe, poverty, both absolute and relative, is not getting better, particularly for many in the developed world. I recognize that the metric for defining poverty differs slightly depending on who is doing the counting and how calculations are made. Even so, while I don't particularly want to get into a debate on the methodological issue, I will make one central point. The unemployed and those in poverty are undercounted in most countries, including the affluent ones. In the affluent societies, there are a large number of people who do not proportionately share the wealth that is being created, and there is a permanent group of people who barely command sufficient resources to sustain life. It can be argued that there are, in most developed and developing societies, signs of a growing permanent underclass. Although the rise of the underclass is not a new phenomenon in the developing world, it has not been a major concern in the affluent West since the Great Depression. The growing number of the marginalized in the developed world is a major reason why increasing social turmoil exists today. It is a social fact that 'middle class societies, as opposed to societies with a middle class are the bedrock of democracy' (Fukuyama, 2014, p. 613). Democracy in the West is experiencing troubled times and the decline in living standards for a goodly portion of the middle class may be one of the reasons why. As the living standards of the middle class shrink, and the wealth of the upper 1% of society increases significantly, right-wing populism, xenophobia, and nationalism have become progressively apparent. Poverty and inequality, in addition to demographic change, are causing great fear and anxiety throughout society.

There is considerable discussion particularly by politicians these days about the middle class. Most view it as the bedrock of society but declining in numbers. But how is the middle class defined? Conventional wisdom would take the economic view and define the middle class in terms of income. From this standpoint, the middle class would earn '0.5 to 1.5 times the median income' (Fukuyama, 2014, p. 609). He goes on to suggest that '(s)ociologists, . . . tend not to look at measures of income but how one's income is earned – occupational status, level of education, and assets' (Fukuyama, 2014, p. 610). This broader definition has the potential to shed greater light on the plight of those who were once part of the middle class, but due to the recent fractionation of labour through de-unionization, increased use of technology in the workplace, and other sociological and geopolitical circumstances, many have fallen out of that social status and become frustrated at their seeming decline. This is particularly true in societies that measure individual worth by the occupation one holds and the amount of consumer goods they command.

There are statistics to suggest that poverty is receding in some developing countries given the rise in their GDP. What is less understood is how much of that new wealth is captured by the rich and how much of it is trickling down to the middle class and the poorer citizens of those countries. It is quite possible for a country's GDP to grow without reducing poverty. In fact, this is what has occurred in many western countries, particularly in the United States (Hedges, 2018) over the recent past.

Much of the poverty discussion does not include the inequality issue but focuses on other economic factors, such as social security rates, job creation, balance of payments, and immigration, legal or otherwise. The governments of capitalist countries seem to view greed as an inevitable and acceptable fact in the capitalist economy. Inequality does not seem to be of great concern in the public policy debate. Granted, the economy should not be considered a zero-sum phenomenon, but when a handful of people own the largest proportion of the wealth of a country, it cannot go unnoticed. Even though the economy has been healthy and grown each year over the last few decades, those earnings have gone largely to the wealthy, and there has been virtually no growth in middle class incomes after factoring in inflation. Much like during the late 19th and early 20th centuries, the question of inequality will need to be considered before poverty and social mobility can be adequately addressed.

The world uses GDP to measure how well countries are progressing economically. Many countries have made progress on this measure despite the gloomy picture presented herein. There is considerable growth in the GDP of some of the poorer countries throughout the world. Many Asian countries have advanced economically over the last few decades. Africa, and some of the other lesser known countries, however, are still mired in searing poverty in comparison to the affluent

West. And, while poverty serves the plutocracy in that it allows for capital accumulation by the top tier of society, it decreases the standard of living for many citizens, increasing tensions in society. This is particularly apparent when we examine the distribution of wealth worldwide. Citing both Branko Milanović (Milanović, 2012), *Global Income Inequality by the Numbers: in History and Now*, and Richard Kersley (Kersley, nd), *Global Wealth Reaches All-Time High*, Bergman tells us that '(t)oday, the richest 8% earn half of all the world's income, and the richest 1% own more than half of all the worlds wealth' (Bregman, 2017, p. 153). Citing a United Nations (United Nations, 2013) publication *A New Global Partnership: Eradicate Poverty and Transform Economics Through Sustainable Development*, he further notes that '(t)he poorest billion people account for just 1% of all consumption; the richest billion 72%' (Bregman, 2017, p. 153). Continuing to highlight these publications, Bregman emphasizes that:

> From an international perspective, the inhabitants of the land of plenty [the USA] aren't merely rich, but filthy rich. A person living at the poverty line in the U.S. belongs to the richest 14% of the world population; someone earning a median wage belongs to the richest 4%. At the very top, the comparisons get even more skewed. In 2009, as the credit crunch was gathering momentum, the employee bonuses paid out by investment bank Goldman Sachs were equal to the combined earnings of the world's 224 million poorest people.
>
> *(Bregman, 2017, p. 153)*

A recent report on poverty in the United States by Alston, the United Nations special rapporteur on extreme poverty and human rights, presents a startling conclusion:

> In September 2017, more than one in every eight Americans were living in poverty (40 million, equal to 12.7% of the population). And almost half of those (18.5 million) were living in deep poverty, with reported family income below one-half of the poverty threshold.
>
> *(Alston, December 15, 2017, p. IV. #14)*

Alston's report puts to rest the notion that extreme poverty is only an issue for the developing world. It is coming increasingly to light that the developed world has its own problems on the poverty and inequality front (see Hedges, 2018).

Even some of those comprising the top 1% are beginning to see the deteriorating situation in the economies of the affluent world. Warren Buffet has been a long-time spokesperson for shining light on this problem. Bill Gates has turned a great deal of his wealth into philanthropic endeavours through the Bill & Melinda Gates Foundation. Buffet has contributed substantial amounts of his wealth to that foundation as well. Notwithstanding this benevolence by a few, the issue is as much a moral one as it is economic. Nick Hanauer is a member of the top 0.01%. Here is what he had to say about himself and those like him in an interview with Catherine Clifford of CNBC in January 2019. She is worth quoting here at some length:

> Hanauer is keenly aware that he is on the fortunate side of America's wealth gap ('Truly, my success is the consequence of spectacular luck, of birth, of circumstance and of timing', he said in 2014 TED Talk); a crisis that he says is 'the direct result' of a 'moral failure'. 'I am a practitioner of capitalism', Hanauer writes in Oxfam's state of wealth inequality report published Monday. 'The most important lesson I have learned from these decades of experience with market capitalism is that morality and justice are the fundamental prerequisites for prosperity and economic growth'. 'Greed is not good', continues Hanauer, who is a signer of The Giving Pledge, an elite philanthropy circle co-founded by Warren Buffett and Bill Gates. 'The problem is that almost every authority figure – from economists to politicians to the media – tells us otherwise. Our current crisis of inequality is the direct result of this moral failure'. In the Oxfam report, 'Public Good or Private Wealth?' which was released as some of the biggest names in finance convened in Davos, Switzerland for the World Economic Forum, Hanauer says the path the world is on will lead to chaos for everyone. 'This exclusive, highly unequal society based on extreme wealth for the few may seem sturdy and inevitable right now, but eventually it will collapse', writes Hanauer. 'Eventually the pitchforks will come out, and the ensuing chaos will not benefit anyone – not wealthy people like me, and not the poorest people who have already been left behind'. Indeed, the inequality Hanauer speaks of is getting worse. The richest 26 people currently own the same wealth as the poorest half of the world, according to the report from charity Oxfam. The collective fortunes of the world's billionaires rose by 12 percent in 2018 – or $2.5 billion each day. At the same time, the wealth of the poorest half of the global population fell by 11 percent last year, Oxfam says.
>
> *(Clifford, 2019)*

Appropriately, Hanauer takes the issue out of being purely an economic problem and squarely defines it as a moral question, which it truly is.

This problem was not always manifest in the United States. Ronald Reagan began decreasing taxes for the rich in 1981. He lowered the top marginal tax rate from 70% to 50% in 1986 and further reduced the rate to 28% before leaving office. Since then the problem of extreme income inequality has grown to where it is today. Using the rhetoric of maintaining competitiveness, other countries followed suit and reduced their rates similarly. These tax measures have produced the inequality problem we witness today in a very short time frame.

The numbers presented herein are startling and expose the extent to which inequality and poverty are inextricably linked and are pervasive throughout the world. Poverty and inequality are both an individual issue and a social problem. The question of poverty and inequality needs to be addressed from both a conceptual and operational point of view. How society defines the phenomenon of poverty, the role poverty-stricken individuals play in its perpetuation or eradication, and the responsibility of the collective in addressing the issue are important matters to be addressed. As you might guess, these questions are answered quite differently by each sovereign country and the worldview of their leaders. Addressing the question conceptually requires a country to determine the definition of citizenship and the extent of poverty acceptable in that conceptualization. Certainly, such international bodies as the United Nations attempt to provide a standard measure of poverty, but societies are so different that it cannot be applied uniformly across the globe, particularly in a capitalist structured world-system.

All societies make a statement about how they view the world and their individual citizens in it. How a country defines and addresses the provision of social policy that affects many of their countrymen, positively or negatively, defines the moral outlook of that society. It may also rely on the resources available to allocate to the problem and where the issue ranks in relation to other priorities. Although many social analysts view present social policy to be entirely inadequate, some feel that a more generous welfare policy would victimize the middle and upper classes given that their tax rate would need to increase if all other expenditures were maintained at present levels. As it turns out, poverty appears to be more acceptable in some regions than in others. The Nordic countries are well known for their intolerance for widespread poverty, while the United States and other western countries are more tolerant of it. As predatory corporate capitalism has become more pronounced over the years, poverty has become further neglected and tolerated, and those suffering from poverty are being increasingly blamed for their situation. In these countries, increased poverty is taken for granted as a component of what it means to be a citizen of a corporate capitalist country. It is seen as an acceptable by-product of the imperative of creating greater profits. How a person does in capitalist society is seen as a function of the individual. If one suffers from poverty it is assumed that it is a result of personal inadequacies they can control. A more empathetic view sees poverty as a social problem and not the total responsibility of the individual.

There are social expenditures in addition to income, such as healthcare, that influence the concept of poverty. For example, in Canada all members of society are provided medical coverage through a single payer system. The single payer is the provincial government with partial funding by the federal government. Health is viewed as a basic right and is one of the major policies that conceptually defines Canadian citizenship. On the other hand, and although the state provides social assistance payments to individuals in poverty, it does not see it as its responsibility to provide a basic standard of living (Guaranteed Basic Income – GBI) as a right of citizenship as it does in the case of healthcare. In fact, there is a constant societal debate about what level of social assistance payments to those in poverty should be assumed by the state and in what form it should be delivered. Guaranteeing a basic standard of living to all citizens in society is thought by many, maybe the majority at this time, to be disincentivizing those who should be seeking employment. Most nations that provide social assistance to their citizens deliver it at such low levels that it hardly brings the recipients up to the poverty line let alone afford them a decent standard of living in comparison to the median level of income enjoyed by the rest of society. Until now at least, no capitalist country has adopted a basic income sufficient for those on it to live a life that reaches the bottom of the middle class standard, although some small countries such as Finland have embarked on an experiment in providing a GBI. There are exceptions to the forgoing statement, however. Many countries, including my own, Canada, provide pensions to their elderly. By most accounts, pensioners in Canada live an adequate life on this pension albeit without any nonessentials or indulgences.

Most countries view social assistance payments as a temporary policy, not as a lifestyle option. So, societies develop social policies to address the marginalized in their midst either through thoughtful discourse or based on the ideological worldview of those in power. The question of poverty is often answered incrementally by operational and policy decisions. Once society has addressed the conceptual question, policies and programs are developed to operationalize that conceptualization.

There are times in history when the human intuition for fairness and equality seems to come to the surface of society providing an atmosphere for change. The confluence of conditions was enough to propel significant changes to the social system when the western world encountered and suffered through a major depression in the late 1920s and early 1930s. Societies from across the world are once again experiencing an increasing separation between the haves and the have-nots. As in earlier times the desire for change on the part of those who are becoming marginalized in society will increase and become progressively more vocal. J. D. Vance in his book *Hillbilly Elegy* chronicles the plight of those left behind by the post-industrial world. He states in the introduction to his book:

> I didn't write this book because I've accomplished something extraordinary. I wrote this book because I've achieved something quite ordinary, which doesn't happen to most kids who grew up like me. You see, I grew up poor, in the Rust Belt, in an Ohio steel town that has been hemorrhaging jobs and hope for as long as I can remember. I have, to put it mildly, a complex relationship with my parents, one of whom has struggled with addiction for nearly my entire life. My grandparents, neither of whom graduated from high school, raised me, and few members of even my extended family attended college. The statistics tell you that kids like me face a grim future – that if they're lucky they'll manage to avoid welfare; and if they're unlucky, they'll die of a heroin overdose, as happened to dozens in my small hometown just last year.
>
> *(Vance, 2016, p. 8)*

What is remarkable is not that Vance's personal story of obtaining a college degree and making it out of poverty is commonplace as he points out, but a story that is becoming more and more unlikely in the industrial, affluent world. I use the word 'affluent' to describe the world in which Vance finds himself, but, unfortunately, that affluence is not being shared across the socio-economic classes. Twenty-first century society is the New Golden Age where the gap between the rich and poorer members of society is being exaggerated. Although growing poverty is clearly an issue for Vance and others like him, it is the lack of any hope that is so utterly compelling in Vance's story. Poverty can be viewed as a short-term situation, but when hope is gone, all is lost. Vance's plight, as he describes it, is making social change inevitable as poverty begins to infect more and more people. It would appear that affluent society is now engaged in a race to the bottom orchestrating the need for social change. Is there a particular point when the slide into hopelessness becomes so unbearable and chronic that change has a chance of being pursued and taking root? I believe the tipping point is here now, but, unfortunately, the situation is generating extreme political and social misadventures such as right-wing populism and nationalism and not positive change that is in the best interests of the population. On the surface, this social trajectory may provide emotional satisfaction to those who are most angry at the ruling class, but not true positive change. Hopefully this negative trend can be reversed and made positive, allowing hope to return.

Fundamentally there are two ways to define poverty. The first is absolute poverty where individuals in that situation barely garner enough calories to stay alive each day. Second is relative poverty where a person measures their ability to consume material goods in relation to the expected consumption patterns for the larger society in which they live. Conceptually, poverty can be observed as an individual problem (social deviance), a social problem (society has not constructed its social expenditures and institutions appropriately), a problem of human exploitation (keeping social payments sufficiently low but adequate to maintain an unemployed pool of labour to keep wages low), or as a political and economic (and, hence, structural) problem. Poverty defined as an individual problem views the unemployed person as a social deviant. In this scenario, society sees poverty as the fault of the individual, not society, and treats it as antisocial behaviour. Poverty viewed as a social problem sees individual poverty in terms of an inadequate social system that leaves individuals disadvantaged and without hope. In this case, poverty is viewed as a phenomenon that is disruptive to social functioning. Poverty can also be viewed as a system of exploitation. For example, the capitalist system, particularly predatory corporate capitalism in modern western societies, defines full employment to include a small (perhaps not small enough) permanent percentage of unemployed persons. If everyone in the labour market was to be employed and there was no surplus labour (unemployed), there would be great pressure to raise wages as competition for workers' increases, putting downward pressure on profits. The capitalist class benefits from a pool of permanently unemployed workers to keep wages in check. Finally, poverty can be defined as a structural political/economic problem. In this scenario, poverty exists because the political system is unwilling to eliminate individual scarcity and redistribute wealth through government policy and programs to those at the lower end of the economic spectrum.

There may be an over-reliance on a romanticized private economic system that assumes that unemployment and poverty are purely temporary and that the natural workings of the capitalist system will eventually eradicate it. That said, we have been waiting for some time for poverty to be eliminated by capitalist ideology and it has not come about. It is time to look for other mechanisms to reach that goal.

Poverty as a structural problem views the condition from an ideological and situational point of view. The replacement of labour by technology in manufacturing and other economic sectors is a new and significant factor in the structure of production requiring a new approach and definition of what it is to be a productive individual. However, it continues to be argued that a basic income or an enhanced social assistance program that provides those without work a decent standard of life would be destructive to individual incentive and creates too much dependence on government support. The best method for alleviating poverty in this worldview is job creation by the private sector. This orientation might be more aspirational than realistic. Many governments have developed programs, such as workfare, to replace social assistance. Workfare programs provide recipients with social assistance payments but attach work-seeking activities contingent to that assistance. Some argue that this practice is tantamount to indentured slavery. Work in this scenario is viewed as what gives meaning and structure to life. This position, however relevant historically, may now be a threat to social solidarity considering the highly anticipated technological future the world faces. Also, the oligarchs who make up the predatory corporate capitalist class can only increase their profits if they continue to put downward pressure on wages or replace labour with technology, so increasing the wellbeing of people either through increasing minimum wages or through public expenditures is not an option for them. Conceptually, if not operationally, a world-system that embraces predatory corporate capitalism cannot effectively address the issue of poverty, although those at the top continue to give lip service to that goal.

Defining poverty operationally is determined by the social policies enacted by various governments. For example, if healthcare is provided by a single payer, usually the government, it suggests that society considers healthcare a fundamental right of citizenship. If, on the other hand, healthcare provision is left to the marketplace, it is deemed by society to be a commodity to be bought and sold on the open market and not a public good. The breadth and depth of social policy provision by a government operationally defines poverty in any particular country. Needless to say, all governments across the world make a statement about how they view poverty by their social policy provision or lack thereof.

Although many governments boast of meagre unemployment rates, their estimates do not account for those who have removed themselves from the labour market because they can't find work, nor does it account for those who are underemployed and making poverty-level wages. Non-participants in the labour market are not counted as unemployed in most countries. Labour non-participation in the economy has been increasing throughout the latter half of the 20th century and continues into the 21st. Much of this non-participation is a direct result of the increased use of technology in the workplace and the lack of skills needed to perform in the new technological economy. This fact may be an unobserved statistic when poverty is being examined by government officials. They prefer to provide the public with good news, and a shrinking unemployment rate is one of those good news stories but only a partial truth. Many who remain in the workforce do not receive regular growth in their income to keep them ahead of increasing inflation. Lack of growth in the standard of living for some is another part of the problem and has only recently become a matter of discourse in the wealth inequality debate. The quality of the work that remains available is never questioned or analyzed. Many workers are in perpetual poverty or on the edge of it, simply because of the lack of systematic increases to the minimum wage bringing it up to reasonable levels. On the other hand, it would be cheaper for employers to move to technologizing the workplace if the minimum wage was increased to appropriate levels. This presents a catch-22 for workers. Many workers at the low end of the income spectrum must work several jobs to make ends meet, and in many cases that doesn't provide them with a decent standard of living.

Notwithstanding the serious issue of insufficient income that is keeping an increasing number of people below or at the poverty line, modern nations do not include social deficits that extend beyond purely financial inadequacies when they define poverty from a conceptual or operational point of view. It is now understood that elements such as health, education, housing, and income constitute one's station in society and their relation to poverty. In a modern world, the ability to participate fully in society is a matter of attaining a basic level of these commodities and services. Today, it should be everyone's right to expect enough of these basic provisions to live a life without constant fear of being poverty stricken or marginalized. In the modern world, poverty can no longer be described simply as the lack of income; there are many other conditions that help define that state (see Reid, 2017).

The notion of poverty is now being replaced, or augmented, with a wellbeing or happiness index. This is a recognition that poverty is not just a matter of income. Many countries, such as Canada, have developed a wellness index to provide a measure that reports how individuals are doing, in addition to, or a replacement for, the GDP measure that is traditionally used for such purposes (see the Canadian Index of Wellbeing – CIW). The GDP measures private consumption, gross investment, government investment, and government spending. The GDP measures economic activity no matter whether it is productive or negative. As you can see, it concerns itself with economic measures only and contains no other indicators. The GDP uses the mean as the measurement statistic, so some people can be doing extremely well, as does a small minority in the United States and elsewhere in the affluent world, and calculate a very positive average, even though multitudes of people do extremely poorly and remain in poverty.

To get a clearer picture of how individuals are doing in society, a wellness index which includes several social variables in addition to the economic measure provides a more comprehensive picture than the GDP. The CIW, for example, includes measures taken on community vitality, democratic engagement, education, environment, health, leisure and culture, living standards, and time use. In addition to individual measurements on each of these categories, an aggregate score is calculated to provide a single measure of how well the reference unit scores overall. The reference unit can be any jurisdiction from the local to the national or international level. Such an index reveals the state of conditions on which the level of social solidarity can be judged.

Social solidarity exists when most members of a society feel the system is working fairly and for their benefit. The idea of social solidarity goes beyond economic security and comprises other elements, including those contained in the CIW. If a minimum level on these variables is not achieved, then social solidarity breaks down, and those who feel disenfranchised will not support the status quo but will work to overthrow the system that oppresses them and move to create a replacement. The recent rise and advance of right-wing populist governments throughout the world is evidence of dissatisfaction among the population with their status in life and the breakdown of social solidarity.

Part, but not all, of the breakdown of social solidarity is the issue of increasing separation between the rich and poor. As Fukuyama tells us, ' . . . societies with extremes of wealth and poverty are susceptible to oligarchic domination or to populist revolution' (Fukuyama, 2014, p. 608). Both conditions can be seen in the affluent western countries today. The upper 1% own 40% of the US economy, and while perhaps not quite as exaggerated, this trend can be seen in most other parts of the developed world as well. Liberal democracy is under attack and suffering duress. The potential of liberal democracy has not been realized by mass society. It has been captured by a minority of greedy people for their own benefit. What is not disclosed to the American people is the fact that the US economy has grown between 2% and 3% annually for the last decade; however, all those gains have gone to the upper economic class. Luce, in his book *The Retreat of Western Liberalism*, tells us that 'the asset value of the world's leading billionaires has risen fivefold since 1988' (Luce, 2017, p. 33). He goes on to explain:

> (s)ince 2009, the US economy has expanded by roughly 2 per cent a year. Yet it took until 2015 for the median income to regain the level it enjoyed before the Great Recession. Perhaps 'enjoyed' is the wrong word. The median income in 2007 was below what it was in 2002, at the start of the business cycle that lasted for most of George W. Bush's presidency.
> *(Luce, 2017, p. 31)*

If Luce's analysis of the wealth gap in the United States and the rest of the western world is not enough to convince us that inequality is a paramount issue to be addressed, Sachs in his book titled *The Price of Civilization* argues that '(t)he wealthiest 1 percent of American households today enjoys a higher total net worth than the bottom 90 percent, and the top 1 percent of income earners receives more pre-tax income than the bottom 50 percent' (Sachs, 2011, p. 22). Worker wages have seen no substantial increase since the late 1980s. It has not been a matter of the rest of the world stealing manufacturing jobs from the United States but the continued inequality that has become structural in the predatory corporate capitalist economy where the problem lays for many people. The major issue that is causing the decline of living standards for the middle and lower economic classes is that the majority of gains resulting from the increase in GDP have gone to the upper 1% for over a decade. Until the real problem is made clear to the public, the predatory capitalist system will continue to experience social breakdown and anomie. And the worst may be yet to come. Robert Frank in a report for CNBC titled *By 2020, half of the world's wealth will be controlled by millionaires*, states:

> (t)he trend will likely accelerate in the next five years as the largest wealth gains are expected to go to those at the very top of the top. U.S. households worth $20 million or more will see their wealth grow by more than 7.6 percent over the next five years, according to the report. Those in the U.S. worth $1 million to $20 million will see their wealth grow 5.7 percent, while those worth less than $1 million will see their wealth grow 1.2 percent.
>
> *(Frank, June 7, 2016)*

These are astonishing statistics. The social system is under great pressure now, so one must wonder what will happen to society if these projections come true. As discussed in other parts of this book, right-wing populism, nationalism, and xenophobia have been on the rise in many western countries over the last decade, leading to the elections of 'strongman' governments as a reaction to inequality, changing demographics, and the breakdown of social solidarity. What will be the consequences if these inequality trends continue? Let's hope that the reaction is positive and constructive. Perhaps it's time to start talking about social democracy as a replacement for our present notion of liberal democracy.

The present frustrations felt by the middle and lower economic classes are being interpreted by right-wing populist governments as the rest of the world taking advantage of them. There is a strong 'Us vs Them' sentiment in many societies across the world (Bremmer, 2018). The globalization system is seen as the locus of social unrest and the target of attack. Strongman governments throughout the world have seized on this unrest and used it to their political advantage. Also, some politicians who have authoritarian aspirations have taken advantage of the COVID-19 pandemic situation to increase their power through legislative action. They also cite the deficit in the balance of trade as a demonstration of what is wrong in the globalized financial world. The present US administration and others also focus on the growing number of immigrants to their country as the problem inhibiting social and economic development. The immigration issue is a diversion from the real problems affecting society today. Increases or decreases in trade with other countries are not the appropriate metrics for determining the root cause of the economic issue, however. No trade with other countries would result in an equal balance of payments, but that is not in anyone's interest. The imbalance in trade has more to do with the huge deficits and debt of the United States and the

American people living beyond their means, not a problem originating externally. The crux of the problem is that the United States is continuing its consumption patterns with the credit card, and the rest of the world is financing that lifestyle in addition to the growing inequality problem. The US governmental debt is also at historical levels. The present US government still sees the repatriation of manufacturing as the saviour of the economy when it should be focussing on establishing the new economy.

What the present administration does not understand is that manufacturing is less important as an economic activity than it was a half century ago. For example, the Chrysler heavy-duty truck plant has announced it is moving back to Michigan from Mexico. When it left the United States to go to Mexico 30 or more years ago, it took thousands of jobs with it. If, in fact, it does return, it will be a plant that employs fewer people because of the new technologies that are now used in auto manufacturing. The jobs in auto manufacturing are also becoming highly skilled unlike when the Chrysler plant left Michigan for Mexico when it employed mostly semi-skilled line workers. The new auto industry requires computer experts to instruct and supervise the robotics, not workers to physically construct the product. Robotics has replaced human labour in the production process. We are moving to something equivalent to 3D printing even in the auto industry. Repatriation is essentially meaningless except for corporation managers and their shareholders who will reap the rewards of the new manufacturing processes. This is the fallacy of the promise to return manufacturing to the United States or to anywhere else in North America. Wages, the present Mexican advantage, is quickly slipping away because of innovations in robotics and the newly signed United States–Mexico–Canada Agreement (USMCA) which promises to increase autoworkers wages in Mexico. When the cost of using robotics is lower than wages paid to workers, manufacturing will gravitate to robotics or to other low wage rate countries. Robotics has another advantage too – robots don't go on strike. Even the most poorly paid worker in Mexico cannot compete with robotics capitalized over a long period of time. It will simply be cheaper over the long-term to move to robotics in manufacturing than to pay workers liveable wages. When this happens, where a manufacturing plant is located will be meaningless except for the taxes it provides to state and local governments. At some point governments will need to think about taxing robots instead of people.

The problems inhibiting social and economic development in other countries is not immigration, as many governments want to suggest, but the failure of governments to deal directly with the internal inadequacies of job growth or wealth distribution and a changing economy. These issues become even more complex as technology invades the workplace. Immigration is just an easy target to blame instead of governmental inadequacy and their seeming inability to think creatively and not rely solely on their long-held ideological view.

Today, intellectual and information services development is perhaps equal or more important to a modern economy than manufacturing. That said, the present US administration is attempting to regress to a 1950s' notion of carbon-based energy production in an attempt to rebuild the modern economy. The attempt to free up the carbon-based energy system is not only consequential for the environment and individual health but entirely inadequate for sustaining the economy over the long-term. What's worse is that the emphasis on the carbon-based energy production and the manufacturing enterprise economy is distracting and may limit the United States' prospects for continuing to lead the world in economic growth. Other countries will grab the mantel and capture the growth provided by the new technologies and the green economy, leaving the United States behind in that sector. If the present US government has its way, it will not lead the world into the green energy revolution but return to a carbon-based industrial system that is unsustainable. The United States is not headed in a direction that will experience the type of economic restructuring that is required to eradicate the issue that has led the country into an unequal and class-based society. The state of poverty will not improve by hanging on to a carbon-based technological society; it can only deteriorate further. Many western countries that championed the idea of equal opportunity in previous times are now heading in the direction of creating a permanent underclass. This problem is not limited to the United States. Conservative governments across the world have a penchant for looking to the past for solutions to social problems rather than looking forward for new approaches. Brexit is another example.

The leaders of society must desist in labelling the problem to be addressed in economic terms. Poverty and inequality today are not solely an economic problem; they are moral questions. Societies throughout the world have the capability to address the issues of poverty and inequality if they so choose. If they choose not to address these issues, then it is clearly a moral failing, and so far, society has failed. What's worse than the past failure to rectify the poverty and the resource distribution problem is the lack of action on this issue today. The suspension of the unequal distribution of wealth discussion, even as inequality grows to astronomical proportions, presents a moral challenge to society. How could this be? The present discourse focuses on the movement of corporations offshore and the advantage taken in trade by other countries as the locus of America's declining middle- and lower-income problem. This is a diversion, one that provides an external enemy rather than focusing on the real moral question which is internal to the capitalist system itself.

Since the beginning of capitalism, the conventional wisdom has been that self-interest leads to social betterment. The present state of inequality would suggest otherwise. Wilkinson and Pickett, in their book titled *The Sprit level*, observe, '(i)nstead of a better society, the only thing almost everyone strives for is to better their own position – as individuals – within the existing society' (Wilkinson & Pickett, 2009, p. 12). They continue in detail:

> The contrast between the material success and social failure of many rich countries is an important signpost. It suggests that, if we are to gain further improvements in the real quality of life, we need to shift attention from

material standards and economic growth to ways of improving the psychological and social wellbeing of whole societies. However, as soon as anything psychological is mentioned, discussion tends to focus almost exclusively on individual remedies and treatments. Political thinking seems to run into the sand.

(Wilkinson & Pickett, 2009, p. 12)

There is no better demonstration of this folly than the recent US administration. The present government has departed demonstrably from past political and rhetorical boundaries and now conducts itself as if it were in a street fight, using racist and misogynistic language to denigrate their critics. This rhetoric is no more than a diversion from the real problems plaguing US society. While the economy is doing well, most of the benefit is not going to those at the bottom of the socioeconomic ladder. We know that the economy will be up and down over time, and how society distributes resources is more important than generating a hot economy that will inevitably cool off at some point.

We have come to the end of an era when growth of the economy is directly tied to individual wellbeing. There is a point where more consumption does not positively affect health and wellbeing but may in fact have a negative effect on the population at large. Obesity may be a prime example of this condition. Obesity is rampant throughout the affluent world where, historically, malnutrition was the focus of greatest concern. A great deal of what we eat today is empty calories, not nutritional food. Unfortunately, cheap food that is available to those suffering from poverty is loaded with calories and not much nutrition. The boundary between want and need is also blurred. Accumulation has replaced many of the other social values, including service to others, as a primary value of society. Our society is rapidly commodifying all resources, including humans themselves.

Economists have clearly enunciated the notion of diminishing returns to personal welfare and the increase in wealth beyond a certain level. The increase in health and overall wellbeing levels off as income rises to the point where there are no further gains in happiness or wellbeing to be had no matter how rich one becomes. In the title of their book, Robert and Edward Skidelsky ask the question, *How Much is Enough?* (Skidelsky & Skidelsky, 2013) Their conclusion to their own question is that we, at least in the affluent world, have far exceeded the aggregate resources required to live a happy and meaningful life. The problem for society, of course, is how those resources are distributed among the population, not their continued growth. Money is no longer simply a means to survival and wellbeing but one of exercising power. It has become the social marker for ranking oneself in society and for exercising control. The 1% use money to control not only those around them but also society at large.

The overwhelming grip on growth in the economy by the 1% at the ongoing expense of the other 99% may eventually cause the citizenry to act and change course. This reaction will be based not only on economic grounds but also on moral considerations given that few people are starving in most of the affluent countries of the world but find the inequality that exists unacceptable and immoral. We have seen in the recent past, fledgling, ad hoc groups of citizens who have risen to protest such imbalance and injustice in society. The Occupy movement and Black Lives Matter (BLM) are such examples, and there will be other grassroots groups that will spring up as time goes on. Although most these groups have been short-lived, they have had considerable impact on identifying and clarifying the extent of the inequality that exists in society today.

Relative poverty is a powerful determinant of anomie and social solidarity, or lack thereof. Individuals perceive their station in life by comparing their situation with others in society. As Wilkinson and Pickett observe when they paraphrase Adam Smith and the theory of capitalism, ' . . . it is important to be able to present oneself creditably in society without the shame and stigma of apparent poverty' (Wilkinson & Pickett, 2009, p. 24). Although absolute poverty may not be as concerning today as it was in Adam Smith's time, the anxiety to which he speaks is still relevant in relation to relative poverty. The conditions of some parts of the inner city appear like Third World in comparison to the more affluent parts of the city which are grandiose and stately. When the stated goal of society is wealth accumulation, crime becomes an acceptable means to achieving that state among some parts of society. The United States has the largest prison population by comparison to any country in the world, and a great proportion of these prisoners are from racial minority communities. If you were to add those in prison to the unemployment statistics, the increase would become more than simply noticeable. One method of keeping the unemployment statistics low is to put more people in prisons. Canada has the same problem. Although the extent of the prison population in Canada is comparable to other countries of its size, the aboriginal population is over-represented in prison. Some northern and remote parts of Canada, in addition to many southern First Nations Communities, mirror the poverty of some developing world economies.

The 20th century also saw the development of gated communities which sent a strong message of exclusivity. The rationale for this development is the increased need for security against increasing crime. This came at a time when crime was on the decrease. Notwithstanding the advertised rationale of security, living in gated communities became a status symbol for many. Where once greed in western society was considered gauche, it is now revered. Corruption permeates our private companies and governments at all levels. Traditional social values are no longer on the surface of everyday life. Moral decay has set in and will alter liberal democracy if not destroy it altogether. A major contributor to this social decay is increasing inequality and the relative and absolute poverty it produces.

There are numerous policies that, if adopted, could significantly increase the wellbeing of the population, particularly those experiencing poverty. First, it is important for policy makers to understand poverty and inequality as a moral, not an economic, issue. The often-heard claim of the lack of affordability of social programs as a reason why we cannot eradicate poverty is not credible. Also, the claim that social welfare disincentivizes the unemployed has not proven to be true. Members of the Democratic Party in the United States produced a manifesto in the form of 'the Green New Deal' (GND) as an aspirational document to guide a discussion on this important subject. The document contains many policy proposals that could have a profound positive effect on poverty in America (see discussion in Chapter 10 for a more detailed account of the GND). In addition to reallocating military budgets for social services, programs and actions such as instituting a guaranteed basic income that is made available to all citizens, recalibrating the marginal tax rates at the top end of the tax structure, instituting free college and university education for all, increasing affordable housing, and growing the provision of mental health services are some of the policies that would affect poverty positively. If adopted by other governments, these same policies could have a very positive effect in other parts of the world as well. For many countries to consider these policies, the affluent world would need to rethink and rework its foreign-aid regimes with an eye to eradicating poverty. This would require a change from using aid as a method for funnelling money to their corporations via the less-developed world. They would also need to increase the size of their support considerably. That said, it is important to consider these and other policies from a systems perspective, not singularly. A general and overarching theme in this book is that instituting policies in one domain will affect life in other domains, so policies for all five of the issue domains discussed in this book must be accomplished in a coordinated fashion and not pursued as isolated events.

Bibliography

Alston, P. 2017, December 15. *United Nations Special Rapporteur on Extreme Poverty and Human Rights*. Washington: United Nations.
Bregman, R. 2017. *Utopia for Realists: How We Can Build the Ideal World [Kobo version]*. Kobo.com.
Bremmer, I. 2018. *Us vs Them: The Failure of Globalism [Kobo version]*. Kobo.com.
Clifford, C. 2019. *Wealth Gap Crisis 'Direct Result' of 'Moral Failure'*. CNBC.
Frank, R. 2016, June 7. *By 2020 Half of the World's Wealth Will Be Controlled by Millionaires*. New York: CNBC.
Fukuyama, F. 2014. *Political Order and Political Decay: From the Industrial Revolution to the Globalization of Democracy [Kobo version]*. Kobo.com.
Hedges, C. 2018. *America: The Farewell Tour [Kobo version]*. Kobo.com.
Kersley, R. nd. *Global Wealth Reaches All-Time High*. Credit Suisse. https://publications.credit suisse.com
Luce, E. 2017. *The Retreat of Western Liberalism [Kobo version]*. Kobo.com.
Milanovic, B. 2012, November. Global Income Inequality by the Numbers. *History and Now, World Bank Policy Research Working Paper*. World Bank, Washington, DC.
Piketty, T. 2014. *Capital in the Twenty-First Century*. Cambridge, Massachusetts: The Belknap Press of Harvard University.
Piketty, T., and Saez, E. 2003. Income Inequality in the United States, 1913–1998. *The Quarterly Journal of Economics* 1–39.
Reid, D. G. 2017. *Social Policy and Planning for the 21st Century: In Search of the Next Great Social Transformation*. London, UK: Routledge.
Sachs, J. D. 2011. *The Price of Civilization: Economics and Ethics after the Fall [Kobo version]*. Kobo.com.

Skidelsky, R., and Skidelsky, E. 2013. *How Much Is Enough: Money and the Good Life [Kobo version]*. Kobo.com.

United Nations. 2013. *A New Global Partnership: Eradicate Poverty and Transform Economies Through Sustainable Development*. United Nations Sustainable Development Knowledge Platform. http://www.un.org/sg/managment/pdfHLP_P2015_Report.pdf

Vance, J. D. 2016. *Hillbilly Elegy: A Memoir of a Family and Culture in Crisis [Kobo version]*. Kobo.com.

Wilkinson, R., and Pickett, K. 2009. *The Spirit Level [Kobo version]*. Kobo.com.

5
GLOBALIZATION
The eye of the storm

Humankind has flourished by advancing across the globe geographically, economically, and technologically since prehistoric times. The latest internationalization of human affairs has spread to the farthest corners of the earth and has had a profound effect on all human life and cultures. The proliferation of globalization is mainly due to increased communications throughout the world. Electronic communication is now instantaneous to all areas of the globe. Corporations have advanced their presence unendingly, to the point where all resources including labour have been commodified and are bought and sold on the international market. Virtually no culture is left untouched by the modern world. What may be different about modern society's version of globalization is its potential to augment, or even replace, the nation state as the primary form of social organization. This potential is producing great fear in the hearts of those who have concentrated their lives solely at the local level and find the prospect of a global social world immensely concerning and alienating. The rise of the new global world has come at a time when other great changes are in the process of altering human life significantly as well. Here I think of the rapid advance of technology into the realm of AI and the pre-eminence of culture over biological evolution in human development (see Chapter 3). These subjects occupy other chapters in this volume, so only a few words on this subject will be offered here. It is important to make the connection between the fast-paced and extensive globalization with which humans are now confronted and the advancing technological world. It is technology that has created a truly global village. Globalization today is mainly corporate globalization. Consequently, predatory corporate capitalism maintains no allegiance to any single country, and although this in itself may not be a bad thing, the lack of corporate ties to place is transforming society across the world.

Globalization is defined as a worldwide commercial and consuming culture.

> The most important role in this was played by capitalism, for global culture is mostly a commercial culture. Next in importance come science and technology, especially the latter for without its technical methods and inventions, most of global culture could not function or be produced.
>
> *(Redner, 2013, p. 517)*

What is missing in this description of globalization is any recognition of other institutions and cultural memes that govern the human community. The definition is solely and simply an economic one. As Appadurai so ably points out, ' . . . economic globalization is today a runaway horse without a rider' (Appadurai, 2001, p. 25). He goes on to explain that in his view,

> (g)lobal capital in its contemporary form is characterized by strategies of predatory mobility (across both time and space) that have vastly compromised the capacities of actors in single locations even to understand, much less to anticipate or resist, these strategies.
>
> *(Appadurai, 2001, p. 25)*

The industrialization of the human production of goods over the course of the 20th century quickened the globalization process to logarithmic speed and has given it pre-eminence in human affairs. The goal of production has changed from focusing purely on survival and human development in the early days of modern humans to that of unfettered growth and wealth accumulation. Pure survival has given way to unadulterated consumption and accumulation for its own sake. Global capitalism is built on the commodification of natural resources, labour, and currency. Nothing on earth seems to escape being commodified. The most recent phase of globalization has not focused solely on populating the unsettled areas of the globe with people, as it did approximately 11,000 years ago when humans first entered North America, and again for a second time by the Vikings in the 15th century, and once again by Europeans in the late 16th and early 17th century, but has now turned its attention to unconstrained material consumption. The latest round of globalization has focused on economic growth facilitated by instant communications and the commodification of all things. The present version of globalization has primarily centred on instant and vast international communication and travel, and the creation of unencumbered markets for trading resources, finished goods, and services around the globe. The pure financialization of the economy has been the recent addition to the globalization project. The transfer of currency to all parts of the globe is now instantaneous given that it can ride the waves of cyberspace. Protectionist trade barriers are continually being replaced by free trade agreements economically connecting all countries around the world. Essentially, the entire world is now one big market with few remaining impediments to it, although there has been a recent hiccup in that process because of a trade skirmish between the United States and China.

The hallmark of liberal democracy has been to promote the globalization of markets and trade. It promotes fundamental human rights and the ideals of equality between and among people across the globe. Liberal democracy embraces the idea of connectedness and is truly internationalist in its outlook. As a result, the world today is more than just a collection of independent states. It has a moral imperative as well. Globalization has attempted to ignite a world conversation about development, human rights, and wellbeing. The bilateral and multilateral trade agreements that have been reached across the world have integrated nation states beyond simply economics to a degree that few thought possible when it all began. That said, some would argue including myself, that globalization has focused too much of its attention on economic growth to the exclusion of many of the other values it purports to promote. If globalism holds dear the notion of increasing human wellbeing and social solidarity across the world it will need to refocus its present effort toward increasing equality and understanding among nations and people. The present form of globalization has failed miserably on that front. What it has done is to integrate production and markets across borders amplifying the financial dependence of each country on others.

The North American Free Trade Agreement (NAFTA) (now USMCA) among the United States, Canada, and Mexico is an interesting case in point. While it was initially intended to be nothing more than an economic pact, it has turned out to be much more than that. Through the creation of NAFTA, the integration of markets between the United States, Canada, and Mexico has created cross-border supply chains that would have been most difficult to unravel if the present USMCA had not been reached. Failure to reconfirm and modernize NAFTA would have inflicted great harm on all countries in the agreement over the short if not the long-term, simply because of the highly integrated manufacturing supply chains that have been constructed since its inception. It would have necessitated the dismantling of tightly integrated cross-border relations and left them in disarray. It is difficult to know just what might have replaced them that could have been equally effective and efficient. There would have been great disruption to those who live in the US states, who depend on the other two countries to receive their products unhindered, if failure to reach an agreement had occurred. The biggest issue underlying the tension in the negotiation was the notion of repatriation of jobs to the United States. But this action would have been costly to consumers and workers who would be asked to pay more for products because of enlarged border tariffs for the goods and services that flow freely across the borders today. As an example, the recent duties imposed by the United States on Canadian softwood lumber increases the price of a house in the United States by approximately $2K to $3K even though the government's rhetoric suggested that it was implemented to protect the American consumer. The only positive benefit to the American economy went to the large US lumber companies and their shareholders who sell their lumber in the US market. It essentially made non-US competitor products costlier and, of course, increased the cost to consumers. It will be interesting to see what effect Brexit has on the prices of goods and services that Britishers will endure once trading agreements are reached between Britain and their eventual trading partners, including the European Union.

The present globalization movement has attempted to move toward a single world economic community while maintaining individual country borders. Although capitalism has always stretched its influence beyond its own borders, the driving force of late-stage capitalism was to connect markets across the globe after the major capitalist countries saturated their traditional markets or when their labour became too expensive. For capitalism to grow, particularly in western democracies, it is necessary for it to spread to lesser developed countries. And, let there be no mistake, capitalism needs to grow continually if profits are to increase unrestrained, and, of course, this is the imperative of capitalism. Polanyi (1944) contends that society is embedded in the capitalist economic system rather than the economy being embedded in the larger social system. It appears that people serve the economy rather than the economy serving people, in Polanyi's view, and I think he has a pretty good take on it.

What is clear is that liberal democracy has been good for the expansion of the capitalist system and continued economic growth, but what is less clear is how it has benefited the working-class citizen throughout the world and particularly for those who reside in western democracies and whose economic life has become stagnant or degraded over the last few decades. In fact, it is becoming clearer every day that globalization, as it is presently constituted, has not benefited the middle or working classes given that wages and standard of living has not increased over the last decade, nor have enough jobs been created to engage the entire working population.

It has not served the developing world well either. And, while some lesser developed countries have increased their GDP over the last while, there is no evidence that this growth has been distributed equitably to the middle class or poorer members of society. What originally were nationally based corporations are now self-serving multinational firms that have become worldwide entities without allegiance to their original or any other homeland. While this may not be a bad thing in itself, there does not appear to be any allegiance to anyone beyond the corporation and the shareholders themselves.

Misperception or not, most people at the beginning of the latest version of globalization thought they would benefit substantially from the corporate globalized system that cast its gaze beyond their borders. But the benefits they were sold have not been realized. Free trade agreements around the world were predicated on the promise of greater prosperity for everyone. The business community promised that all would benefit by expanding trade across the globe, and while owners of capital have profited handsomely from globalization, workers have not. Even the workers in low-wage countries who have found new jobs continue to be mired in poverty because of the substandard wages paid by global corporations, saying nothing about the deficient working conditions they endure. Global companies have effectively de-unionized the world's workforce, and corporations are now in complete control of the world's economy. Because of these dynamics, the working and middle classes have recently demonstrated their dissatisfaction with the concept of globalization. They believe 'the globalists' have sold them out, and who can blame them. Careful analysis might suggest that globalization built on liberal democratic ideals has focused more on the economic aspects of liberal democracy than on the human rights and equality values that were once a very prominent part of the ideology. Neoliberalism has jettisoned the social and human rights values that were part of the original liberal democratic promise. As a result, to many working people now, change means 'loss' that is to be avoided at all costs. In their assessment, many of the earlier changes facilitating globalization need to be rolled back, not enhanced. There is an increasing call by many for deglobalization and economic nationalism. The increasing disenchantment we witness among the marginalized in liberal democratic society today is challenging the very utility of liberal democracy in the minds of many. In some countries, liberal democracy is being transformed into illiberal democracy. Giroux tells us that 'according to Orban {Prime Minister of Hungary} the state is not defined by democratic values, but by its economic and cultural interests, interests that cohere among a growing number of far-right regimes' (Giroux, 2018, p. 181). The focus is purely on power and economic interests at the expense, in many cases, of human rights, worker safety, equality, and cooperation, both internally and externally. This reaction to the marginalization of large portions of the population has promoted the rise of right-wing populism and authoritarian governments, a subject discussed at great length elsewhere in this book.

There are many ways of examining and identifying the various factions in the present globalization debate. David Goodhart (2017) provides an instructive lens for examining and describing globalization and the various sectors of the population that are affected by it. Primarily he sees the great divide as comprised of those in favour of an 'open' system and those who view the world through a 'closed' lens. He sees these perspectives replacing the right and left axis that has defined modern politics in democratic countries around the world. In his book titled *The Road to Somewhere: The Populist Revolt and the Future of Politics*, he suggests it comes down to the 'Anywheres' and the 'Somewheres' people. The Anywheres people are university educated and likely employed in the technology or finance sector. They are people who are engaged in, and benefit from, the globalized economy. They see a future use for their skills. They can be regarded as the achievement society. They can live anywhere and are prepared to do so. Anywheres people make up 20% to 25% of the population of Britain, according to Goodhart. Of course, there are no estimates of the size of this group in the United States or elsewhere around the globe, but we can speculate that it is similar.

In contrast to the Anywheres group, half of the population of Britain is thought to comprise the Somewheres class. These are people who are not graduates of college and see the good-paying jobs they once occupied leaving the country, or more likely, being eliminated by technology. They are pushing back against liberal democracy and its enthusiasm for globalization. They subscribe to the 'change is loss' viewpoint. The Somewheres people identify solely with their communities and neighbourhoods and look inward and not outward. They are often fuelled by fear and anger, as they are increasingly marginalized by neoliberalism and the globalized society. Goodhart has some sympathy for the Somewheres people when he states, '(w)ithout a more rooted, emotionally intelligent Liberal Democracy that can find the common ground between Anywheres and Somewheres, the possibility of even more unpleasant backlashes cannot be completely ruled out. . .' (Goodhart, 2017, p. 21). An example of the backlash Goodhart is referring to is Brexit, the campaign and eventual vote by Britain to leave the European Union. The rest of the population is seen by Goodhart as 'Inbetweeners', sometimes agreeing with the Anywheres and at other times with the Somewheres' point of view.

I have no doubt that Goodhart is right when he suggests that the answer to this problem lies in finding the common ground between the two groups and then moving to it. The question, of course, is finding the common ground to which all can feel some level of comfort and then designing a process to move toward it. The size of the common ground may be very thin in comparison to the two extremes, however. Politics today seems to be hyperpolarized, and there is no clear path to bipartisanship in elected assemblies or in society at large. People live in a communications bubble and listen and communicate with only those people who share their point of view (Bremmer, 2018). Dialogue between and among a diversity of views is virtually non-existent. The move to wide-ranging dialogue and eventual agreement among the majority of the populous will take great skill and leadership. It will also take a great leap in imagination and goodwill on the part of the political leadership and the public at large. Unfortunately, there doesn't seem to be a lot of goodwill around these days that could support such dialogue.

But before moving to designing processes that will lead there, it is necessary to define the problem in detail and not simply jump to one tribal view of the future as we seemingly have done with the Brexit vote and the election of Donald Trump to the presidency of the United States. It is extremely important to see how we have gotten off track. How has globalization failed a large segment of the population? Or, how can we generate a new approach to creating a different form of globalization and construct something that will benefit all of society? While many paths to a new world-system will need to be considered, there is one clear necessity in this quest and that is the need to reach agreement on the most pressing areas of concern and a clear enunciation of the collective future aspirations of the population. No longer can politicians continue to sell their constituents a fantasy about going back to the good old days where they are independent from the rest of the world and better off for it – or, conversely, attempt to continue to provide a false promise of good days to come under the present structure of globalization that is slanted toward the predatory corporate capitalist system. In my view, neither of these approaches is tenable, as the difficulty of implementing Brexit and its apparent early consequences are demonstrating.

A key consequence of the most recent form of globalization has seen the shift in labour-intensive manufacturing from West to East. This shift has had the consequence of changing the makeup and face of labour in both eastern and western countries. Essentially, high-paying manufacturing jobs have been transferred from Europe and North America to the far eastern countries of the world where labour is cheaper and environmental laws lest strict. In exchange, the West has created a 'new economy' that emphasizes technology, and technological information creation and analysis, which essentially replaces labour in the production process. The new economy has produced a sophisticated financial services sector that continually strives to create new financial instruments and craft additional ways of manipulating financial transactions. This shift is not too dissimilar from the transition from an agriculture-based economy on which Europe and North America depended in their early years, to a focus on manufacturing during and after World War II. The present transition requires the conversion of a highly mechanical manufacturing system to one based on digital and AI technology and high-end services such as finance. The new technological and service jobs require a high level of education that all members of society do not necessarily possess. It also appears that society is not able to produce as many high-paying jobs in the new industries as have been lost in the manufacturing sector, although that fact is not acknowledged by capitalist rhetoric.

At the low end of the jobs spectrum in the new economy are service jobs where the less-educated members of society provide personal service to those at the top end of the financial ladder. Usually, but not entirely, these service sector jobs are low paying in comparison to information producing activity and financial sector employment. The low-end jobs have engaged many of those who previously earned higher wages in the manufacturing economy before that sector migrated offshore or automated. As you might expect, the new jobs shedding economy has produced great dissatisfaction among those who have been hurt by the change. Because of this shift in the economy, western countries have produced a structural unemployment rate greater than during the heyday of the manufacturing economy. And, while unemployment rates seem low in some countries at the moment, more revealing is the labour participation rate which tends to be retreating in most developed countries. Many of those permanently removed from the world of market-based work are the residual of highly paid manufacturing jobs that were once available to those without high levels of education and now without the necessary skills to participate in the new high-skill economy. The basic economic snapshot presented here earlier is the fundamental reason there has been a growing discontent and anxiety among a large sector of the population, most notably those who once held high-paying manufacturing jobs but are now stuck in the low-income category, if they are employed at all. They see themselves as victims of globalization, whether that is reality or not.

The established political class is suspected of negotiating trade deals with other countries that are responsible for the transference of good jobs offshore. It is also suspected that global corporations are replacing permanent citizens with immigrants who are taking their jobs for less pay. This has caused considerable backlash and loss of confidence in our political class. The anti-establishment mentality leans toward replacing the political elite with a group of aspiring, myopic, political wannabes who tend to be xenophobic and nationalistic and call for isolationist and protectionist measures without understanding the ultimate costs of such policies. There does not appear to be any statesmanship in politics today, only the rhetoric directed at going back to a more familiar and comfortable bygone era. This 'circle the wagons' mentality often prefers an authoritarian leadership, one that leans closer to the far-right end of the political spectrum than the mainstream politician. Perhaps the greatest illlustrations of the outcomes of this unorthodox political phenomenon are Brexit in the United Kingdom, the Le Pen ultra-right-wing movement in France, and other ultra-right-wing leaders in Europe.

The crusade that Donald Trump created in the 2016 US election has been self-described as a movement rather than a political campaign, even though he has attached himself to the Republican Party. Some might argue that Trump is trying to change the face of democracy and not simply working to improve it. There is often a yearning for a more fundamentalist law-and-order regime and a return to an earlier version of society when confronted with such momentous changes to the social system as globalization. One of the main characteristics of a populist movement is a return to isolationism and nationalism and a rejection of full and active participation in geopolitics.

The momentous change produced by globalism and internationalism is particularly difficult for those who feel more comfortable in the structure of the past. The structure to which they identify doesn't exist any longer, nor can it be retrieved. The world has moved on from that point in time, and the present social and technological environment would not support such a return to those earlier days. Those caught in the transition to the new economy blame their diminished status on the globalization phenomena and all that it entails. And there is no doubt about it, there are many people, mostly the undereducated and under-skilled, who are being hurt by globalization as it is now practiced. But unfortunately for some, returning to the 'good old days' is not possible because the social conditions to support those days no longer exist, and the genie of internationalizing the economic and technological revolution cannot be put back into the bottle. Globalization is a fact, and all societies, particularly in the developed world, will need to deal with it. The new technology that is now deeply ingrained in our culture demands it. The task at hand is to reinvent the social structure, including establishing a new raison d'être for many individuals in it. Gaining purpose in life beyond the work environment needs to be the focus in the future. Sense of individual purpose will need to be found in a social environment that does not require all people to work in the traditional sense. To me, this is the crux of the transition facing society, even though some of our politicians are still fighting the 'return of manufacturing jobs' rear-guard action. Many believe that the first step in returning to the 'good old days' is to reject any form of globalization.

Richard Haass (2015) explains that there is a nuanced but important difference between an international system and an international society. The world-system of which he speaks is a globalized system constructed through a scientific reality that incorporates advanced technology, instant communication, and AI, which is truly revolutionizing our everyday lives beyond anything we have experienced to date. The competitive international system in which we all live is constructed by a number of independent state governments that are trying to construct rules and norms on which to operate an international exchange system effectively and efficiently, and what may be needed is to reconstruct a system that benefits everyone, not just the few. The first step in this process is to acknowledge that the international system is now a driving force in human life and truly in control of our day-to-day lives. It may be time for nation states to construct a replacement system with the introduction of a cooperative structure that can more effectively manage the advances in technology and communications the world has created and continues to advance. This suggests a movement from an international system toward an internationally focused and centred society. Extra-national rules and regulations addressing the forces that affect all nation states, whether they occur within a nation's boundary or not, need to be created and implemented by an international society. It is becoming increasingly clear that many of the problems faced by humanity are not boundary specific, global warming being the prime example. The human cultural system will need to evolve to incorporate and manage the new technologies if it wants to maintain control of them. The world community needs to further embrace globalization and not shrink from it, as the world may seemingly want to do today. Instead of constructing the system from the purview of the nation state, we should be designing something that extends the internationalism of the system in order to enhance and uplift all people throughout the world. We need to create policies that will take all countries and their citizens to the next level of security and prosperity, no matter how that is defined, and those strategies need to focus on wellbeing not just on economics as expressed through the GDP measure. The zero-sum game, where there are winners and losers, needs to be abandoned. The goal of the new globalization must be to enhance individual life not the wealth of corporations and their shareholders. No longer can we buy the rhetoric of companies that if they do well, so will the individual in society. This has not proven to be true in the past and should not be part of the conventional wisdom of the future.

In his book, Haass also suggests that sovereign countries can do pretty much whatever they desire, within limits of course. The difficulty with that notion is that what countries do in this new technological and information age significantly affects other countries in the international system, whether intended or not. As a case in point, we can examine the present US administration's effect on the environment as it dismantles the Environmental Protection Agency. The United States is opening up environmentally sensitive areas to development without regard for the environmental costs to the planet. These policy decisions affect the entire world given that pollution knows no boundaries. Perhaps a greater example is the US administration's withdrawal from the Paris Climate Accord. Such policies made by one government have implications for the entire world's population, not just their own, so the international order must be collectively governed particularly when it comes to such areas as climate change. No government is truly sovereign in such matters even though they may want to be or sometimes act like they are.

Systems theory focuses on the connections and relations between and among parts of the system rather than reducing and examining the separate functions and their underlying properties. Systems theory examines the relationships among the parts of the system and how each part changes in relation to the others when those other parts are stimulated or changed in some way by external or internal forces. As we reorganize the international society, we must concentrate on the relations among countries, shortening the distance between them through closer ties and more substantial rules of engagement. At this moment, the countries of the world seem to be moving in the opposite direction and they appear to be taking up the drawbridge and retreating into their insular worlds. This is a kneejerk reaction to the globalization of the world that has been designed to benefit corporations and the rich and powerful members of society. It is a result of focusing on changing some parts in the system and overlooking the impact of those changes on the other parts. It is no wonder that those who have been adversely affected by the globalized system are fighting back against an arrangement that is bent on disenfranchising them. What is needed now is to make major adjustments to the system and the relations among the parts so that all benefit and not just those in positions of power and who have been in control of the system to date. Rather than retreating behind borders in order to preserve the present state of inequality, we need to rethink what is in the best interests of humankind and the environment it depends on daily. I recognize this is a tall order, but the continuance of human society may well depend on it.

To address the fallout from globalization and the anxieties it has produced, new thinking and strategies that may initially appear to be outside the current box of emotional and rational thought are required. To adequately address the situation described in this chapter, a leap of creative imagination by all humanity and their governing systems is urgently required. The consequence of not addressing these issues is too great for them to be left unresolved.

Does globalization have to maintain its present form – where there are clearly winners and losers and a dramatically skewed system in favour of a small group in society – to survive? It should be clear by now that I don't think so. Appadurai clearly believes there is an alternative to the present system and who should be in control of globalization. When speaking of globalization, he suggests:

> [A] series of social forms has emerged to contest, interrogate, and reverse these developments and to create forms of knowledge transfer and social mobilization that proceed independently of the actions of corporate capital and the nation-state system (and its international affiliates and guarantors). These social forms rely on strategies, visions, and horizons for globalization on behalf of the poor that can be characterized as 'grassroots globalization' or, put in a slightly different way, as 'globalization from below'.
>
> *(Appadurai, 2001, p. 11)*

This depiction of globalization appears to embrace a wider notion than the present one. This altered form of globalization may display its essence through the idea of 'internationalism' which connotes an integration beyond just the economic world and embraces ideas about laws and social justice principles, such as human rights.

Appadurai goes on to suggest that creating a truly international society will not happen without successful efforts to organize from the bottom up. Creating local grassroots institutions and then connecting them regionally and internationally will be the foundation on which this new social order can be successfully constructed. These institutions will rely on goals and objectives that favour the people they are intended to serve and not the international predatory corporate capitalists and elites of society who now dominate the landscape. It is no longer acceptable for the present power structure and international institutions to create what they believe to be appropriate goals and then attempt to sell those goals to the rest of society. The idea of development under a philosophy of 'trickle down' has not shown itself credible in the past. The bottom-up approach to change must replace the top-down approach of yesteryear.

To progress to a sustainable globalized social contract, education, both formal and informal, dedicated to encouraging those at the local level to understand more fully their own fundamental needs and how to organize them into a plan of action, is essential. Many plans of this nature have failed in the past because the elites of society, including many well-meaning professionals, have short-circuited this stage of development because it is time consuming and deals with what some might consider to be abstract concepts rather than concrete actions. Often, well-meaning community and political organizers know the outcome they want before they engage the wider public in the process and simply try to sell their ideas rather than listening intently and responding to the needs and aspirations of the community. Their approach seems to be more in the camp of marketing than in community development. Concrete actions can come later but certainly not before local needs are clearly identified and articulated at the grassroots level by those the eventual actions are designed to benefit. While this process may seem slower, it builds sustainable support for actions leading to long-term solutions. Only through a dedicated action like the one enunciated a little earlier, which places the needs of the people at the head of the priority list, will a truly grassroots globalization from below result. Not until those who have been traditionally left out of the globalizing effort are made preeminent in it can there be a satisfactory globalization project that meets the needs of the people. Globalization that benefits everyone and one that achieves a sustainable environment have been left out of the present edition of the globalization movement, and we now see where that has taken us.

The failure of the old system is not only economic but also spiritual. The liberal democratic ideals of reaching out and engendering true social development and progress throughout the world needs to be re-established. The notion that we are in this world together and that cooperation is the path to long-term sustainability and not engagement in a zero-sum competition needs to be reasserted. At the present time, we seem to be acting out of suspicion of global connections and institutions. Changing this attitude toward the organization of the world's affairs cannot be done simply by arguing the economic case but must include discussions on culture and individual purpose in life. Secular meaning in life must intensify and religious fundamentalism decline. Religious fundamentalism, both Christian and Muslim, is now part of the problem that pits one society against another. The merits of maintaining the liberal democratic tradition of open and cross-border conversation versus the right-wing populist ideas of closed societies competing with one another needs to be clarified. Politics, both national and international, as it pertains to globalization, is about culture and not only about power politics. Inevitably, globalization must be perpetuated from the bottom up and not from the top as in the present version.

The major factors that influence and control the process of globalization from below are also encompassed within a temporal and changing environment, both physical and social. It is necessary to examine the 'on the ground conditions' that will eventually push globalization from the bottom. While there may be too many of these factors to examine exhaustively in a single chapter, it is important to focus on those that will produce the greatest effect on globalization from below.

Much of the disparity and inequality in wealth between and among countries and their populations has been produced by those in charge of the globalizing project. Corporations and their wealthy shareholders have constructed a globalization that has left the common person out of the benefits when they should have been included as true and equal partners. The present structure of globalization has simply replaced colonization and feudalism carried out by nation states during the 18th and 19th centuries, with huge international corporations seeking resources and profits from abroad. In fact, some (Hedges, 2018) have argued that a new form of feudalism is being practiced by global corporations in the developed world as well as in less-developed countries. Much of the present globalizing project has included paying graft to local politicians in developing countries who then make decisions that are not in the best interest of their constituents but in favour of those paying and receiving the largess. The only thing that is different between the colonization of the 18th and 19th centuries and the effort of globalization today is that corporations have replaced the nation state as the perpetrator exploiting the resources and culture of the colonized. The other major difference today, unlike earlier times, is the modern communication systems which constantly reflect the life of the affluent to those who are without. This creates expectations that cannot be fulfilled. Globalization is also a function of creating new markets when older markets have become saturated and cannot produce continued high rates of growth and profits, a constant imperative of capitalism. It is also a function of seeking natural resources that are necessary for the smooth running of the capitalist economy. While some give lip service to lifting up those in poverty, reality would not support their rhetoric of benevolence. And, while small progress may accrue to the marginalized in the lesser developed countries, their improvement is infinitesimal in comparison to the gain realized by the globalists.

What people in the affluent countries must come to understand is that the resolution of the migration issue is not to erect higher and stronger border walls in order to guard the homeland but to solve the problems of war and poverty in those parts of the world that are suffering greatly and from which the refugees are fleeing in great numbers. Bottom-up globalization could help with this project. It would promote the merging of local people with business leaders and philanthropic organizations as equal partners in wealth-producing activities at the local level. Focus would be on all areas of social and environmental policy and not just economics. The developed world needs to assist the underdeveloped countries of the world establish stable and functioning governments if they have any hope of stemming the growing immigration they face at their borders. Helping other countries develop and institute sustainable governments does not mean creating institutions that look like the West but resemble indigenous forms of governance that have meaning to local people.

Unnatural country boundaries result in the creation of rivalries and often strife. African countries for example, possess boundaries that were created by colonial powers in an earlier era and do not reflect any natural or indigenous criteria such as language or traditional culture. It so happens that many of the refugees referred to earlier are from African states. New boundaries, if boundaries are needed at all, must be negotiated in the future if more congenial borders and nations are to be sustained. Not until then will bottom-up globalization have a chance of taking root in dire circumstances such as those found in Africa. Local social organization in these societies is required so they can participate fully in bottom-up globalization planning. In fact, establishing local social organization structures may be a necessary first step in preparing the groundwork for launching comprehensive participation in the 'new' global project. Part of this initial groundwork will be to embrace the notion of self-determination and self-government. Nation building may need to be placed back on the agenda. Self-determination and self-government as considered here does not embrace isolationism and extreme competition that leads to conflict but the establishment of a socially confident self-identity that allows societies to seek betterment in a cooperative rather than competitive or conflictual manner. The process needs to begin with the notion that all societies in the negotiation will be better off after a locally attentive globalized system is established rather than the unequal outcomes emanating from the top-down globalization model we have created to date. Bottom-up globalization will depend on equality among the participants in the relationship rather than the unequal arrangements found in the present system. The inequality in the present arrangement is causing great distress and encouraging much of the turmoil we witness today in many parts of the world. In fact, it is fair to say that the status quo is being rejected by many people across the globe because of the absence of basic structures at the local level and the disregard for local culture and the welfare of the people by the present system. Globalization should not be regarded as a zero-sum game but a process where all partners in the relationship gain and advance toward the goals collectively established. These few fundamental principles provide the basic underpinning on which globalizing from below can proceed and be successful in the end. Also, the international community needs to establish human rights standards and embed them within the globalization project.

Continued exploitation of weaker societies for the sole benefit of the exploiting nation or corporation would not fit into the new approach of bottom-up globalization. Globalization will only work if people see benefit to them in the relationship. Until now globalization has been built on the interests of commerce and specifically the large multinational corporations. More specifically, globalization has been constructed by the developed world through exploitation of the countries of the less developed world, although large numbers of people at the lower economic levels in the affluent countries are starting to feel its effects on them now as well. Much of this exploitation has occurred because capitalism in the West has saturated their own markets and now needs to expand into non-saturated markets of the less developed world. This expansion has been accomplished by the exercise of monetary power whereby the affluent countries control the international financial institutions such as the World Bank (WB) and International Monetary Fund (IMF). These institutions lend money to underdeveloped countries who often purchase goods produced by the affluent countries. As affluent countries come under financial stress, they often grant foreign aid on condition that the money is used to purchase manufactured goods from the donor country. This is what transactional agreements in international affairs has come to mean. What this approach does is make the developing world more dependent and more in debt. The indebtedness this has created reinforces the power relationship, thereby maintaining the industrial world's oppression of the less developed countries.

Globalization has promoted the transference of manufacturing to the less developed world were labour is much cheaper and environmental laws less strict, if there are any laws at all. This form of exploitation reduces manufacturing costs and drives down wages in the developed world as well. Labour in the industrialized countries cannot compete with the developing world simply on wages alone. The movement of manufacturing offshore by the West is one of the primary reasons there is an isolationist movement in many parts of Europe and particularly in the United States. It is anticipated by many that this nationalist movement will return jobs to the home country, but this may be simply false hope in so far as many of those jobs have been replaced by robotics and no longer exist even in the developing world. Many of the remaining jobs only exist because of extremely low wages, and if repatriated back to the West they would soon be robotized and disappear completely.

In addition to the economic imperative of globalization the religious force is once again expressing itself. This negative expression is often referred to as culture wars. Many politicians in the United States have been heard to say that closing the borders to immigration is nothing short of the need to maintain civilization. This can also be seen symbolically by the forceful reintroduction by the present administration of wishing others 'Merry Christmas' rather than the more inclusive greeting of 'Happy Holidays'. This may seem like a trivial gesture, but it is of symbolic significance to many as is the phrase 'Make America Great Again'. This notion of making America great again is, as stated before, about making America White and Christian again.

The same may also be said about the Muslim world. ISIS is all about creating a Muslim caliphate. The caliphate if successful would impose strict Sharia Law as an organizing principle rather than the more liberal democratic notion of the rule of law based on a secular constitution. While these differing but similar aspirations would on the surface appear to be isolationist and exclusionary, let there be no mistake both of these religions ultimately desire to export their belief system across the globe. In the case of the West, this has already occurred, and if the United States was still preeminent in that process there would be no notions of isolationism. The aspiration of isolationism by many western countries is a result of the loss of their hegemonic position in the world. The examples above contain the ingredient of religious fundamentalism. Extreme Islamists such as ISIS have founded their movement on an incorrect interpretation of the doctrine that defines the Muslim faith. Many of those leading the isolationist movement in the United States bind their views to a revisionist idea of the far-right Christian religion with many out-and-out false ideas or, at best, misinterpretations (Hedges, 2006).

As hinted throughout this chapter, it appears that deglobalization is now being championed by populist movements throughout the world that would prefer the isolationism and nationalism that populism promotes. There are even some who are asking the question; is this present rise in right-wing populism the beginning or continued decline of the West, or simply a brief moment in its transformation? I think the jury is still out on that question. The answer may rest on whether or not western societies can embrace the good things about the present edition of globalization because there are some, and remedy the problems that have taken the project off track. Central to this discussion of the globalizing question will be society's ability to include people who have been excluded to date. This may seem like a simple task, but it may prove to be more difficult than first thought. It will mean sharing power not only within each of our societies but also between societies, and until now, this has not been something the powerful have been willing to do. It will take great movement on the part of the wealthy to realize their long-term welfare depends on a greater share of the benefits of globalization going to those below the 1% level. All citizens of the world have a great stake in the environmental and social health of the planet and, therefore, must have a greater voice in shaping the globalization project. The desires and aspirations as well as the spiritual life of those left out of the project today must take greater precedence in decision making and in matters that ultimately affect their lives. As Habermas (1983) termed it, the life-world must be better balanced with the system-world (see discussion in Chapters 9 and 10) if there is to be a rebalancing of the social and environmental fabric and include all communities in governing the world and not just the select few. At present, our societies are over weighted toward the system-world and this, in large part, is what is producing the rise of right-wing populism and the rejection of the global system and liberal democracy.

In moving in the direction cited above, society must strengthen the arguments for the democratic society – maybe social democracy – which stresses world-wide communication and cooperation not pure competition, and stresses greater equality and social justice that focuses on the welfare of all citizens. This does not include the long-held, but totally discredited trickledown theory of economics, but embraces the role of government in bringing equality and social justice to all citizens. It is an old cliché but true nonetheless, that citizens must be considered just that – citizens – and not simply consumers. The new globalizing project while maintaining focus on economic health must also strengthen its pursuit of social justice and social solidarity. This is instrumental for rebalancing the life-world with the system-world. Populism will not result in satisfying the legitimate desires of those who embrace it even though it may provide some short-term emotional satisfaction. As alluded to earlier, whether you support populism or liberal democracy and globalism, the population breaks down on this issue mostly by demographic characteristics. As Goodhart summarizes when speaking about the British population, '(t)he richer and better educated you are, the more global your attachments are likely to be. The rich and well educated are also less dependent on national social contracts, while many in the bottom 50 to 60 per cent of the income spectrum have become more dependent on national social contracts in recent years' (Goodhart, 2017, p. 131).

The blending of liberal democracy and the basic principles of social democracy has the potential to deal with the crisis in which we now find ourselves and the basic dissatisfactions of those who feel marginalized in society. While social democracy will maintain the need for modern technological societies to interact globally it will also put the community (life-world) above the corporation (system-world) in that process. If whatever is being proposed at any given moment is not clearly seen to be in the best interest of the life-world but is tilted toward a special interest group or the system-world, the proposal must be considered suspect. All proposals and social structures, including economic structures, must benefit all parties affected and not seen as slanted toward one of the players in the system. No longer can any country be solely driven for its own benefit and at the expense of all others. 'America First' is no longer an appropriate idea when single country policies have considerable repercussions around the world and affect the lives of many people living in other regions. Whether we like it or not, today's technology will not allow us to live in a closed society. We live in an open society and the question is, just how open do we want to make it? Perhaps, more precisely, how open can we make society in a globalized system and still feel comfortably balanced within our national identity? I think we are only beginning to explore that discussion and many governments are drawing very stark margins on which the conversation is beginning to take shape. Unfortunately, some of the right-wing populist movements include a hefty dose of racism which is rejected by a clear majority of people, but this type of thinking taints the populist project in the eyes of most citizens. In addition to racism and xenophobia, right-wing populism embraces nationalism and isolationism which is no longer possible in today's world. Fortunately, technology and other social forces will demand that we move forward with the international integration project or humanity as we know it could slide into another Dark Age. Fear of cultural annihilation through a changing world, perceived or otherwise, produces these xenophobic and populist notions. As a result of the anxiety caused by the uneven development of globalization there is now great backlash against the continued internationalization of the economy. Capital moves very rapidly throughout the world and can be deployed quickly in countries with lax labour laws and non-existent environmental regulations. The middle and lower economic classes are at a point of demanding change. Unfortunately, the response by governments has been to point to false causes for the malaise and to deflect away from taking responsibility for the present, negative, state of affairs. Globalization provides the convenient target for the difficulties being experienced. Right-wing populism and nationalism are the answers for some to the dilemma in which we find ourselves today, but they will not lead to a better world.

We may be about to experience a brief period of deglobalization as governments introduce extensive tariffs on trade and implement non-tariff barriers and sanctions on other nations. Deglobalization will be particularly punishing in the case of economies with long established effectively functioning cross border supply chains. These relationships produce products from materials and production processes that stretch across numerous countries and not from materials or labour found solely in any single region. Today it is very difficult, if not impossible, to find an automobile that has been built in a single country, for example. Deglobalization is also a result of the struggle against the shift in economic and political power from West to East. China is the particular target of the United States for trade tariffs (and with some justification) as China becomes the economic centre of the Asian region. China is also challenging the United States for world supremacy in economic and geopolitical affairs. Likewise, Russia, while punching above its weight, is also challenging US hegemony. This is a political struggle felt across the world.

Capitalism is built on cutthroat competition that has created a race to the bottom both internally to countries as well as internationally. The world-system is changing, and the question remains; what values will guide that change? Certainly, the highly competitive capitalist economic system will need to be replaced with a system that values cooperation over competition and an arrangement that values pluralism over a rigid singular existence, one that benefits everyone not just those with wealth and power.

So, as the saying goes, where do we go from here? Certainly, I don't see a full-bore adoption of the nationalist project by the American people or anyone else for that matter. As mentioned earlier in this chapter and elsewhere in this book, the new technologies will not allow for that kind of inward retreat. The new generation of young people are so addicted to their smart phones and other pieces of modern technology that the nationalist project seems to be out of the question.

It seems that the most likely alternative is for the global community to do a reset of globalism and construct a new form of international society, to use Haass's term (see Haass, 2015). This will come in a form of globalization that not only respects main street more than it does now but places their needs higher up in the priority system than at present. It will also need to be much more respectful of the physical environment on which life depends so that the economy and values inherent in the economic arrangement work for all people rather than viewing humans as simply inputs in the economic system. While this may sound utopian, we are becoming such a powerful species on the planet that we will have to be much more mindful of our activities or we run the risk of destroying ourselves and eradicating many other species in the process. In effect we may make the planet uninhabitable. Eisenstein reiterates the pessimism that emanates from those who study the perils of humanity when he states, '. . . it becomes abundantly obvious that we are doomed. Politics, finance, energy education, health care, and most importantly the ecosystem are heading toward near-certain collapse' (Eisenstein, 2007, p. 551). Let's hope he is wrong and remind ourselves that the frontier mentality where nature needs to be overcome, and everything is taken by brute force, is long behind us. It is now time for mindfulness regarding both the planet and the people on it.

Although our present form of globalism has not necessarily been friendly to the general population of the world but largely served the elite and wealthy, thinking and acting globally from a broader perspective is critical given how the natural and social worlds transcend the artificial borders constructed centuries ago. The operation of the world is truly global so we must act globally, albeit from the perspective of the life-world not the system-world (see Chapter 9). The seeming rejection of thinking and acting globally is a kneejerk reaction to the corporate hijacking of international relations over the 20th century but creating a new world-system will allow humanity to develop a different course for fashioning international relationships for the future. The present chaotic state of world affairs will eventually cause sober thought to prevail once we get over the nationalist attitude that exits now.

Humans have faced crises before and have thoughtfully, or sometimes not too thoughtfully, overcome those challenges. I believe that we will do so once again. The nationalism, isolationism, and reckless disregard for the environment, and the culture wars that we witness today, are a wakeup call to the issues that humanity must address. Hopefully, we are at the beginning of an important worldwide conversation about the future of human development on this planet and not in the final stages of destroying it. A bottom-up globalization must be a major part of the reconstitution of the new world-system.

Bibliography

Appadurai, A. 2001. Grassroots Globalization and the Research Imagination. In *Globalization*, edited by A. Appadurai, 10–38 *[Kobo version]* Kobo.com.
Bremmer, I. 2018. *Us vs Them: The Failure of Globalism [Kobo version]*. Kobo.com.
Eisenstein, C. 2007. *The Ascent of Humanity: Civilization and the Human Sense of Self [Kobo version]*. Kobo.com.
Giroux, H. A. 2018. *American Nightmare [Kobo version]*. Kobo.com.
Goodhart, D. 2017. *The Road to Somewhere: The Populist Revolt and the Future of Politics [Kobo version]*. Kobo.com.
Haass, R. 2015. *A World in Disarray: American Foreign Policy and the Crisis of the World Order [Kobo version]*. Kobo.com.
Habermas, J. 1983. *The Theory of Communicative Action*. Boston: Beacon Press.
Hedges, C. 2006. *American Fascists: The Christian Right and the War on America [Kobo version]*. Kobo.com.
Hedges, C. 2018. *America: The Farewell Tour [Kobo version]*. Kobo.com.
Polanyi, K. 1944. *The Great Transformation: The Political and Economic Origins of Our Time*. Boston: Beacon Press.
Redner, H. 2013. *Beyond Civilization: Society, Culture, and the Individual in the Age of Globalization [Kobo version]*. Kobo.com.

6
MIGRATION
The clash of civilizations

The migration of people from depressed and war-torn countries has been a major issue for not only those countries directly affected by the chaos, but also the jurisdictions where refuge is sought. Much of the present migration is coming from the Middle East and Africa. A great deal of this migration is from Muslim countries that have a very different culture than many of the receiving host countries, which exacerbates the problem. Also, there is great concern on the part of some North Americans about the increase in migration coming north from South America. Migration and the associated problems that accompany it is now such a large issue for some countries it has become a major focus for the United Nations (UN). The UN has issued a fundamental announcement on the situation in a statement titled the *Compact of Refugees Declaration*.

> Record-breaking numbers of refugees and migrants are moving across international borders, fleeing conflict, persecution, poverty and other life-threatening situations, or responding to labour and skill shortages and demographic changes and seeking better opportunities elsewhere. Their journeys can be fraught with peril; appalling tales of tragedies feature daily in the headlines. Those that make it to a destination are frequently met with hostility and intolerance. Those host communities making an effort to provide relief are often unprepared and overburdened by the sheer numbers arriving. Responsibilities are not well distributed: a small number of countries and host communities host disproportionate numbers of refugees, asylum seekers and migrants. Beyond loss of life, the large displacement of populations has broader implications for the social, economic, and political landscape. The international response needs to be more robust and collaborative amongst a variety of stakeholders to address large movements of refugees and migrants. The UN system, NGOs and partners are all working to highlight

the issue, to secure commitments for assistance and to strengthen the collective response to the crisis.

(United Nations, 2013, p. 1)

Perhaps the greatest factor producing migration for a large number of individuals is the turmoil and strife throughout many parts of the world. People are leaving certain countries in droves to preserve life and limb and in search of a better life without poverty. Migration is one of the main factors that is causing those in the affluent countries to retrench into a dark form of nationalism and xenophobia. Not only is much of the migration we see today a result of the conflict and violence within home countries, but also the conflict between countries. In addition to civil war and unrest, global predatory corporate capitalism is a main contributor to poverty in many of the less developed countries that are experiencing out migration. One of the world's best kept secrets is the size of the international arms trade that perpetuates human dislocation. This huge and unaccountable business is exploiting and profiting from the vulnerable in society. Admittedly the arms trade is a symptom of a larger problem, it is global in nature and responsible for much of the turmoil in the world. The international community needs to address this death producing industry in addition to the other factors contributing to forced migration.

Over the last decade or two, war and economic deprivation have become a driving force for increasing unrest in the developing world. More recently, migration is driven by global warming that is changing the habitability of large parts of the globe. Soon we are likely to experience an increase in climate refugees. Large numbers of people are prepared to risk life and limb to escape eternal poverty and violence in their home country in the search for safer lands. Many are forced to flee their homelands because of catastrophes produced by severe weather. These factors are driving many people to seek more tranquil ground. As global warming becomes increasingly felt across the world, people are moving from lands that are no longer habitable seeking something more liveable. Permanent drought or continual catastrophic weather occurrences make migration to new lands imperative. This emerging phenomenon demonstrates the interrelatedness between the major problems of global warming, increased violence, and migration. Until the first two problems are addressed immigration and its contingent issues will remain unresolved and perhaps get worse.

Many migrants seek settlement in western democracies. As a result, we are witnessing cultural tensions in both the United States and European Union. The traditional population in these countries is extremely worried about the Muslim religion taking over and dominating social life, including the replacement of the traditional legal system with Sharia Law. As unfounded as this notion is, talk radio and other more extreme ultra-right media outlets preach this message to their listeners, often and loudly. They see society in terms of winners and losers and argue it is up to the losers to make their way in the world without relying on government aid, particularly when much of the burden falls on western democracies.

Although arguably racist, many of those unnerved by immigration reason that their established place in society is about to be overwhelmed by foreign cultures. There is a struggle going on between the ultra-right and the liberal left in many western societies and the immigration issue provides one of the flashpoints for that conflict.

Because of the swelling numbers of migrants seeking asylum and the cultural differences they bring with them, there has been an increasing resistance to immigration by the population of the affluent countries that have been selected for potential resettlement. The refugee issue is now so volatile that reducing this stress and smoothing resettlement is high on the United Nations' (UN) priority list. The turmoil migration is causing human societies across the globe is now provoking a heightened political debate within and among nations. The UN compact that all member countries are being asked to adopt spells out their concern and aspirations for the treatment of immigrants in the potential receiving countries.

> This Global Compact is the product of an unprecedented review of evidence and data gathered during an open, transparent and inclusive process. We shared our realities and heard diverse voices, enriching and shaping our common understanding of this complex phenomenon. We learned that migration is a defining feature of our globalized world, connecting societies within and across all regions, making us all countries of origin, transit and destination. We recognize that there is a continuous need for international efforts to strengthen our knowledge and analysis of migration, as shared understandings will improve policies that unlock the potential of sustainable development for all. We must collect and disseminate quality data. We must ensure that current and potential migrants are fully informed about their rights, obligations and options for safe, orderly and regular migration, and are aware of the risks of irregular migration. We also must provide all our citizens with access to objective, evidence-based, clear information about the benefits and challenges of migration, with a view to dispelling misleading narratives that generate negative perceptions of migrants
>
> *(United Nations, 2013, p. 2)*

The need for this compact arose from the growing resistance to immigrants, in some cases violence, in the host countries. Some countries, such as the United States, have taken draconian steps, some would say inhuman and racist actions, to stem the flow of immigrants to their homeland. We only need to take a close look at the border between the United States and Mexico to see the tension. Other countries, such as Hungary and other Eastern European states, are taking equally draconian measures to shut out refugees. There are such strong feelings about immigration in the United States a noticeable proportion of the population is calling for a wall to be built on their southern border despite the evidence that it is not an effective or efficient solution to the issue. The current president of the United States is the wall's greatest advocate.

The resistance to asylum seekers and immigration is not exclusive to the United States. The German population resisted Chancellor Merkel's admittance of up to a million refugees during the Syrian crisis. Some would say the migration issue has effectively put an end to Merkel's chancellorship, however her approval ratings have recovered somewhat for her handling of the COVID-19 pandemic. Resistance in Hungary led by their prime minister is also noteworthy as an example of the rejection of immigrants coming out of Africa and the Middle East. Italy has also displayed similar feelings. France has endured anti-immigrant marches in the streets and a great deal of Brexit in Britain has to do with stemming the flow of immigrants across their border. France and Quebec in Canada have banned certain types of religious dress in various circumstances, demonstrating their unease with the cultural practices of newcomers.

Why such resistance to immigration? In addition to the fear of the 'other', the idea that new immigrants will take jobs away from long-term citizens, the threat of terrorism, and the seemingly strange cultural practices of newcomers, play a large role in the unease. The advent of the COVID-19 virus has demonstrated that racism and anti-immigrant sentiment still exist in our societies. Many Asians whether they are first generation or fourth are being verbally, and in some cases physically, attacked and blamed for bringing the virus to their communities whether true or not. It would appear that it doesn't matter where they have been, but who they are.

Discontent among the unemployed and low paid workers in the West manifests itself in various ways. Some of these manifestations are dark. New immigrants and those who enter a country illegally are often fingered as the source of all trouble in society, particularly any increase in crime. This is particularly the case when those immigrants, legal or otherwise, are from an extremely different culture that displays unfamiliar customs or ways of behaving and what appears to many as fundamentalist religious practices and beliefs that are at odds with the host culture. This is an old refrain, and likely the ancient fears of those not like us was ingrained in the human condition when we were hunters and gatherers. It appears that we continue to maintain our deep-seated sense of territoriality. We have always feared or been suspicious of the other. And, while this attitude may have helped us survive in those perilous times and in that harsh environment, it is now inhibiting our progress as a species. The populist politicians take advantage of the disquiet and stoke those fears as a way of gaining or maintaining power. Immigrants make an easy but unjust target on which to attribute the problems confronting modern society. Much of the present round of migration is from diverse parts of the world that practice different and, to some, strange cultures than previous migrations of people to the West. The issue of culture may be at the heart of the unease with migration. Misunderstood cultural practices can have profound negative impacts on those who do not comprehend their nuances.

Culture has always played a predominant role in human life. Culture encodes the perceptions and actions of a large group of people and helps them conceptualize how to explain or conduct everyday life. It also provides a narrative for how life began and for the afterlife. Following the dawn of humankind ancient humans became aware that life was finite and precarious so began to create explanations for an afterlife that transcends existence here on earth. They reasoned that much of their daily success depends on supernatural events or the good graces of several deities as well. Life events were not entirely in their hands and they needed the assistance of forces beyond their control to survive. Nature and fertility were often the focus of the mystical world. Separate Gods watched over the birth and death cycle as well as providing abundant harvests and the animals used for food and clothing. Human self-awareness gave rise to traditions, group practices and, most importantly, religion, that sustained everyday life through difficult times. Early on, religion took the form of multiple Gods each with a specific function, and then much later Gods that resided beyond this planet were created to oversee all of life's contingencies. Each religion has their own God who directs their lives through unique cultural practices. Different tribes and groups of people defend the supremacy of their God, including engaging in violent conflict. Conquered peoples often faced the acceptance of the conqueror's deity or were put to death.

As new religions came into existence established practices resisted the new beliefs and tribes often carried out bloody battles in hopes of exterminating the new forms of worship. As a prime example, Christianity was under attack by the Jews and the Roman pagans during its infancy, as has been well documented. This competition between groups for the 'true path' has been ongoing throughout the history of humankind and it continues today. It is often religion and culture, as well as language, that separate the various nations throughout the world. In the early beginning of humankind, people could not leave these everyday life-and-death affairs to chance. They needed help in navigating the world and the Gods were there to assist them. During the Axial Age around 500 BC, the great religions and philosophies replaced the 'nature Gods' as the great societies emerged during that period. The prodigious religions of Christianity, Islam, Hinduism, Buddhism and other minor religious practices are an extension of the early beliefs. It has been mused that many of the stories found in the Christian Bible originate from the basic narratives created by earlier religions, often pagan religions. In addition to providing contact with the supernatural world, religion and culture also provide societies with a hierarchical structure for social organization. Individuals identify strongly with their particular culture, it provides them with rules and a framework for making sense of the world and for carrying out the day to day affairs of life. The fundamental beliefs encoded in culture and religion may be at the heart of the mistrust of the immigrant.

Whatever the nuanced form religious practice, takes it has played a central role in cultural development worldwide. There are very few if any peoples in this world where religion does not play a central role in their specific culture and moral outlook on life, even though many nations profess to be secular states. Religion has been central to guiding human life down through the ages. Much of the suspicion of the 'other' is rooted in the different religious and cultural practices that collide. These practices may appear to confront or be in contradiction to the cultural practices of the host country even though there may be no basis in fact for that judgement. Much of the fear rests on suspicion and ignorance. The notion of an infiltration of Islamic terrorists with the intent of carrying out clandestine actions against the host country finds no basis in fact. Those who are attempting to migrate are purely seeking a better life for the most part or fleeing a desperate, violent, situation in their home country. A great deal of the violence in society is committed by home grown domestic terrorists not recent immigrants.

Although religion provides guidelines for how one should live everyday life and worship the deity, it has not always been a positive force in the lives of humans. Early in modern human life, and here I speak of Cro-Magnon (10,000 or more years ago) or perhaps even further back, religion anointed special people known as shamans and priests who specialized in interpreting the nuanced teachings of that culture's religion and who claimed special access to their particular God. It was the priests and shamans who organized and led society as it formed and carried out that people's way of life. Priests were often the jurists and officials who enforced the rituals and taboos within the culture. It is the priest who communicated directly with God and interpreted His teachings to society. Although perhaps a little more sophisticated today, the influence of religion and culture on the mores and beliefs of societies throughout the world is considerable. Many of these mores no longer fit the time and, therefore, may be detrimental to the needs of present-day society. The idea of the Rapture in some forms of conservative Christian religion for example, proposes that there will be a point in time when the earth will be totally destroyed and true believers will be transported to heaven on that day. Many of the adherents to this belief reason that global warming or other forms of environmental collapse don't really matter because the world will be destroyed in the end anyway and they will be transcended and saved from the chaos. In fact, many of these people want to hasten the event. They view global warming as one of the possible first signs of the coming Rapture. Many religions view other religious practitioners as infidels and enemies of their culture.

Cultures have often been chauvinistic viewing their fundamental practices as the only true way to live life and their God as the only true God. Because of this chauvinism, wars have been fought, and many people have been slaughtered in the defence of religion and the culture that contains it. In modern times, the clash has been particularly evident between the Christians and Muslims, between Muslims and Jews, and between Islam and Hinduism. Intra-sect conflict within religions is also evident today. This conflict is evidenced by the ongoing struggle in Afghanistan between the fundamentalists and the more liberal moderns. Within Islam there has always been tensions including outbreaks of violence between the Sunni and the Shiite sects. Certainly, many of the wars fought in the Middle East and in other regions around the world are carried out in the name of religion. Conflicts over religion occur between religions and between factions within religions. Apparently, religion is something worth fighting over or at least thought to be by ancient and modern humans.

Religious conflict was also evident in India when the country eventually split into two separate nations. The Hindu population remained in India while the

Muslims created a new country next door called Pakistan. Although the great Mahatma Gandhi and Nehru attempted to keep the Indian subcontinent one country, the pull of Islamic culture and religion eventually tore the country apart and the new nation of Pakistan was created in 1947. It was essentially set aside for the exodus of the Indian Muslim population and the separation of the Hindu from the Islamic culture. After great struggle it was determined that each religion needed its own territory and governance structure. Regardless of that determination the two countries who exist side by side continue to live under tension at their border even after the partition. Open conflict in Kashmir flares up from time to time because that part of the territory was never settled and remains in dispute to this day. The exodus of European Jews from Europe to Palestine after World War II is another example of separation by religion. There has been unending conflict between Israel and the Palestinians ever since. The Christian Crusades in the Middle East and North Africa are prime examples of the intense conflict between cultures. Muslims, or Moors as they were known at that time, and the Christians fought long and enduring wars in order to capture the holy sites claimed by both religions. These wars were also fought in the name of God and to annihilate what was thought to be an illegitimate, rival, culture and religion. The 'other' is often known as the infidel or unbeliever. The word Crusade and Infidel to this day maintain powerful negative connotations in both cultures. Anti-immigrant feelings have had a long and tortuous history. The strength of those feelings ebbs and flows but today these tensions seem to be on the rise and gathering strength as a political issue.

One cannot dismiss the exercise of raw power when it comes to inter group conflict and relations. That said, it is difficult to separate out the role of religion in society from the pure exercise of power. They are often inseparable. Culture and religion have provided the framework through which power is often exercised. Religions control their adherents through coercion that is attributed to a supernatural deity who hands down strict rules and guidelines on which to conduct the life of the faithful. There are dire consequences including the threat of becoming a social outcast rained down upon those who do not conform to the 'Word' and the directives set out by those in control. In some cases, noncompliance can result in death.

Adherents to a religion do not have a choice in conforming to the dictates of the faith in question. If the axioms of the faith are not followed then dire consequences are predicted, perhaps the greatest being sentenced to living throughout eternity in Hell or its equivalent. Faith provides convenient supernatural explanations for naturally occurring or human induced events. There were many of the Christian faith who believe that the destruction of the Twin Towers in New York City on 9/11 was an act of God to punish the United States for its failure to strictly comply to His word. Those who carried out the events of that day believed they were doing the work of Allah. Religion gets tangled into politics and overlays the complexities of immigration and its role in society even though western countries profess to separate church from state.

In a classic work titled *Jihad vs. McWorld*, Benjamin Barber (2001) provides an analysis of the long running story of conflict between two cultures. Jihad and McWorld can be understood as euphemisms for Islam and Christianity although the conflict has taken on a tone of modernity vs. traditional culture, in this case as expressed by a strict, some would argue, rogue, form of Islam. And, while the conflict has taken on religious overtones the struggle is also about the rejection of modernism and the preservation of tradition. Although the West, particularly those who occupy the far-right political spectrum, view Islam with suspicion, it must be pointed out that Islam also sees many of the western practices as confrontational to its own religious existence and, more importantly, to Allah. Cartoonists have been murdered by radical Islamists because they were thought to blaspheme Allah through their drawings. It is important to note here that moderate Islam does not recognize terrorists such as ISIS as legitimate representatives of their faith or their notions of world order. In fact, the Muslim majority speak in opposition to the values expressed by this new movement. Regardless, it is a clash of cultures as the title of Barber's book suggests, and he points out that the modern version of the conflict resumed openly on September 11, 2001. Here is what Barber says on the subject:

> On September 11, Jihad's long war against McWorld culminated in a fearsome unprecedented and altogether astonishing assault on the temple of free enterprise in New York City and the cathedral of American military might in Washington. In bringing down the twin towers of the World Trade Center and destroying a section of the Pentagon with diabolically contrived human bombs, Jihadic warriors reversed the momentum in the struggle between Jihad and McWorld, writing a new page in an ongoing story.
>
> *(Barber, 2001, p. xi)*

It is important to note that Barber has identified the infrastructure under attack in religious terms when he refers to the World Trade Center as a temple and the Pentagon as a cathedral. I don't believe the use of religious terms to describe the attack was an accident but very deliberate.

Culture is very personal and a strong mechanism for creating social solidarity that gives identity to the individual and the group in which that individual resides. That said, culture is also dynamic and not static. If cultures don't change in accordance with changes to the social and physical environment, they die out. The world and cultures within it change over time and the cultural artefacts that societies produce can have profound effects, as we have come to know through such diabolical instruments as the creation of nuclear armaments. And, while cultures can change over time they do so slowly and, arguably, fall behind the rate of technological change occurring in society. One of the difficult transitions in human existence has been the accommodation of new technology and social practices while not altogether abandoning fundamental beliefs. This psychological and sociological conflict is exacerbated during times like the present when human existence appears to be engaged in a great technological leap forward that requires adaptation of new cultural practices at a quickening speed but appears unable to do so institutionally and culturally. Many of these changes require relinquishing some long-held traditions such as the unrelenting drive for economic growth at the expense of the environment. Culture helps us navigate the world in which we live, and it becomes very difficult to give up formerly successful practices for new ones that have not yet proven themselves but seem appropriate. This process becomes even more difficult when individuals leave one culture and enter a new one as happens when families migrate to new countries that are guided by a very different set of cultural practices. The inhabitants of the host culture often view the newcomer with suspicion and with great anxiety, but immigrants often display great unease with many of the new culture's practices. This may be one reason why all countries demand that the newcomer take an oath of allegiance when being given citizenship.

There are some additional underlying tensions that promote cultural clashes throughout the world today. These clashes are found within, as well as between, nation states. An imbalance in resource use and wealth creation among the countries of the world, where some exist at a much higher economic level than many of the others, is certainly one of those tensions. It has been roughly estimated that Europe and North America contain 20% of the world's population but uses approximately 80% of the world's resources. The West has dominated the global economy since the end of World War I. The signing of the Treaty of Versailles at the conclusion of World War I is significant in that the United States, France, and Britain carved up the resource rich parts of the world particularly in the oil rich Middle East, creating new countries that did not exist before the war. Many of these countries were artificially configured with virtually no interior coherence. They also distributed each of these lesser states among themselves providing them with colonies or quasi colonies.

Russia morphed into the Soviet Union taking over many parts of eastern Europe and the Baltic states after World War II. These colonies provided the raw materials that allowed the colonizing countries to develop economically while keeping the colonized states poor, giving expression to the centre/periphery theory of development. The historic domination of the super-powers over the peripheral countries is beginning to show signs of unravelling as part of the new global social economic transition begins to take shape. This disentanglement is a major part of the tensions permeating western society today. The countries that have decolonized, particularly in the Middle East and Africa, and where political and social struggle seem to be high, are the countries that contribute large numbers of immigrants to Europe and to other parts of the world, including a smaller number to North America.

North America is encountering a different set of circumstances and experiences than Europe as it relates to migration. Much of the immigration confronting North America, particularly the United States, is a result of extreme poverty and the many conflicts that are occurring in South America (SA). It should be noted that much of the violence occurring in SA can be contributed directly to the neglect of these countries by the United States. After years of attempting to manipulate or prop up many of the SA governments friendly to the United States but not necessarily to their own people, the United States has recently reduced or stopped providing support altogether. The isolationist policies of the present US regime appear to many to be an abandonment of the SA people. Also, the neglect of SA countries by the United States has encouraged the election of some unsavoury governments in this region. The corrupt practices of many SA governments are leading to violent conflict and causing poverty among the citizenry of those countries, and this neglect and deteriorating conditions in the homeland constitute a major factor for the large numbers of migrants attempting to enter the United States. Most, if not all, of the SA migrants eventually appear at the border between Mexico and the United States. I must point out that this migration from Mexico to the United States is not as large as the migration from the Middle East and Africa to the European continent. Notwithstanding that fact, the US administration has taken extreme measures to block migrants from entering the United States from Middle Eastern Muslim countries so the threat of travel from that region to the United States has been somewhat curtailed. The US administration has also limited the time that migrants from the Caribbean who were subjected to displacement because of a series of natural disasters can remain in the country. Many are now in the process of being deported back to their homeland. As a result, some of those displaced persons who do not want to go back to their original homeland are attempting to cross the northern US border into Canada and are seeking refugee status there. This influx of asylum seekers has put extra pressure on the Canadian immigration system leading to some consternation by a small but growing minority of Canadian citizens. As a result, Canada is now experiencing a backlash to immigrants from other cultures and some politicians are using this fear as a political weapon. It is fair to suggest that immigration has now entered the political arena in North America with more force than it probably deserves given its actual impact on the traditional cultures of those countries. The 'other' continues to be viewed with suspicion and resentment, however.

It is also clear that other parts of the globe are challenging the present economic domination of the West. China, and to a lesser extent India, are such rivals. The pressure on cultures to change the imbalance in consumption of resources across the world is a major factor for much of the unrest that exists today. This tension is expressed by the nationalism we witness in many societies across the globe particularly in the European Union and United States. This tension gets expressed by anti-immigration rhetoric and policies. Even as the economic imbalance shifts and some of the Asian countries increase their standard of living, the affluent countries react to this trend with xenophobia and nationalism.

The 'Make America Great Again' slogan and the introduction of punishing tariffs on China, Europe, and other countries throughout the world is how the United States has greeted the rise of the East and the decline of the West. There is a sense of entitlement in the United States. Many believe they should be allowed to continue their economic hegemony without challenge. Some of the affluent countries of the West appear to believe it is their divine right to exploit the rest of the world. The United States, as expressed elsewhere in this volume, maintains a huge deficit and debt that is financed by many of the countries whose population has been mired in poverty for many years. Interestingly enough, China and Saudi Arabia service much of that debt. Individual citizens in the affluent world have been living on their credit for the last decade in order to maintain their standard of living. The US government is now emulating that practice. The deepening of the US debt began long before, and apart from, the COVID-19 pandemic, however, it has increased significantly because of measures taken to cope with the disease.

The tensions surrounding immigration in many of the affluent western countries is part of the clash between the modern and traditional cultures of the world as they intermingle and, perhaps more importantly, as the economically emerging countries challenge those in decline. This challenge is exacerbated by the increase in sheer numbers and the demographic projections based on the birth rates for each of the cultural groups inhabiting the developed world countries. The birth rates of the non-European population in many western countries will soon dramatically change the demographic makeup of those countries. This fact is giving rise to white nationalism in many of these countries and particularly in the United States and some eastern European nations.

While some nations continue to live beyond their means through exploiting the resources and cheap labour of the lesser developed countries, many of the developing countries find themselves facing unbearable poverty and violence. During colonization by western governments many colonial states were not only deprived of their sovereign right to development but learned the art of corruption from their colonial masters. Despite a number of these nations gaining their sovereignty after years of colonialism they have maintained the artificial boundaries set by their predecessors and, in many cases, have created corrupt governments in emulation of their former colonial masters. As a result, they suffer from internal and external rivalries and conflicts that continue to tear them apart leading to increased migration by their inhabitants. Many of those residing in war-torn and poor countries migrate to other more promising lands using dangerous methods, many at their peril. Countless individuals have lost their lives on this journey.

Perhaps most telling are the birth rates in many poor countries. Although birth rates in the affluent countries have been on the decline over the recent past, the poorer countries of the world have experienced an increase in their population. The high-birth-rate countries cannot create a sufficiently robust economy to supply jobs fast enough to absorb the increasing population, ramping up pressure on migration. Complicating the immigration issue further is who the immigrants are in terms of their ethnicity and cultural practices. Unlike previous high migratory times to the New World, the new immigrant is of a different race and religion in comparison to the predominant group in the receiving country. As a result, the affluent countries of the West are not only experiencing unprecedented levels of immigration than they have for some time, but the immigrants are from very different cultures engendering greater impact than what might be expected as normal. This is likely to intensify given that the world's population increase is projected to come from Asia, Africa, and the Middle East in the future. The practices within these cultures are not understood or appreciated but frightening to the host country and their traditional population leading to uneasiness, strife and, in some cases, violent backlash.

While mass migration is nothing new, it comes at a time when the traditionally affluent countries of the West are in decline because of the issues raised earlier, whereas in former times it was clear that migration, particularly to North America, was necessary to build the new world. Until approximately the middle of the 20th century the United States had a somewhat lax immigration policy and maintained open borders. Today, it seems to many that immigration is in some way detrimental to the continued prosperity of the affluent countries, the destination targets of many migrants. What is in fact the case is that the economy of many western countries needs immigrants to maintain their population and the robustness of their economies in the face of declining birth rates among the long-time citizens.

Although it may not be true, many citizens of the destination countries feel that the new arrivals are taking jobs away from them. As a result, xenophobia and racism are now on the rise in many of the immigrant receiving countries. That is not to say that racism and xenophobia did not exist in many of these countries historically because it certainly did. Notwithstanding that fact, racism is raising its ugly head again after a time when it was politically incorrect to display such distasteful behaviour. There seems to be some idea that a country can close its borders to immigrants and continue to flourish and, in some cases, regain what is perceived to have been lost. Some including myself, would argue that immigrants, particularly asylum seekers, are the new scapegoat for a country that may be in decline or otherwise struggling to maintain its hegemonic position in the world, but not the real problem.

It is a known but often forgotten fact that the population of the affluent countries of the West is getting older and declining demographically. The birth rates of the industrial world are decreasing, and natural population replacement is not outpacing death rates fast enough to maintain growth in the population. Consequently, affluent societies have come to depend on migrants to bolster their declining numbers. Piketty provides an historical perspective to this dilemma.

> Demographic growth accelerated considerably after 1700, with average growth rates on the order of 0.4 percent per year in the eighteenth century and 0.6 percent in the nineteenth. Europe (including its American offshoot) experienced its most rapid demographic growth between 1700 and 1923, only to see the process reverse in the twentieth century: the rate of growth of the European population fell by half, to 0.4 percent, in the period 1913–2012, compared with 0.8 percent between 1820–1913. Here we see the phenomenon known as the demographic transition: the continual increase of life expectancy is no longer enough to compensate for the falling birthrate, and the pace of population growth slowly reverts to a lower level.
>
> *(Piketty, 2014, p. 103)*

So, whether society wishes to increase or reduce its dependence on immigration to maintain the economy or not, it will need to take in more people if it wishes to continue or maintain its present population balance. Although many argue that immigrants will be needed to replace the present workforce as it retires, it is more likely the case that it needs them to be the new consumers and, perhaps more importantly, those who will provide new energy to the culture as the present population ages. This will occur as automation continues to replace people in the production process. However, the scenario of continued growth depends on maintaining the capitalist model that may be unfeasible for reasons discussed elsewhere in this book.

Despite the well documented demographic transition enunciated by Piketty above, the political climate during the first few decades of the 21st century focuses on the negativity of immigration. Elections in Europe and North America are being fought on immigration issues where before the economy was the number one question. The current US president made immigration a large part of his platform, including the promise to build a wall on the southern border to eliminate what he calls illegal immigrants coming in from Mexico. It needs to be pointed out however, that many of these asylum seekers crossing the Mexican border are from other South American countries and not solely Mexican citizens. Many of the farmers in California and other surrounding states have relied heavily on these so-called illegal immigrants to harvest their crops for many years. As part of this anti-immigrant movement some of the western countries that prided themselves in being liberal democratic, including the United States, are leaning more toward authoritarian governments than they have over the recent past, maybe ever, in part because of this immigration issue.

Some of the former Soviet countries, such as Hungary and Poland, that have been recently on the path to becoming liberal democratic and members of the European Union appear to be reversing course and have elected authoritarian leaning governments largely because of the immigration issue. Italy, a long-time member of the European Union, has also recently elected a nationalist government, elected on an anti-immigrant platform. Turkey, another county that is seeking full admittance to the European Union, is also moving toward an authoritarian government, if it isn't already there. It is clear, that the social transition to a new economic and technological world is creating many pressure points and tensions, one of which is the backlash to immigration.

The form this immigration backlash is producing, is the embracing of tribalism and nationalism. Although historically new immigrants often settled in enclaves in the larger cities throughout the world there was an overall feeling of unity and social solidarity within those countries. Certainly, there has always been backlash to immigrants but not to the extent that it is today, and covertly or not so covertly, sanctioned by government. Historically, policies of integration, assimilation, or multiculturalism were enacted by governments to accommodate the new arrivals into the population. Governments tried to speak for and to, the issues of a variety of people and did not govern solely for those who elected them. This may no longer be the case. Some governments see themselves as representing a specific community with an explicit dogma, and it sees it as their duty to defend that community from what has become known as the 'other'. Tribes are now forming in countries that were once the leaders of the liberal democratic ideal that promoted inclusion not division. The 'other' in most cases, is the immigrant or, more accurately, those who do not share the physical or social features of the mainstream culture or religion. Even when the minority is the majority, governments that are elected because of their connection to the owners of extraordinary resources govern to protect themselves and those that look and behave like them. This pattern is problematic when it becomes clear that what are now the minorities will soon become the majority in society given the low birth rate among the long-established tribe and the need for more immigrants to do the work that the mainstream citizenry can't or don't want to do.

There are some policy and programmatic actions that are available and can help lessen the backlash to immigrants and immigration. Most of these policies and programs are designed to lower the rhetoric and temperature in society by reducing the fear of the 'other'. Perhaps the most basic action to be taken has to do with the basis on which immigrants and different cultures are received by the host culture. This involves understanding that culture provides the lens through which individuals make sense of their world. Present models such as the melting pot theory of integration are wishful thinking but not realistic. To think, for example, that a newcomer to a country is likely to abandon their way of seeing the world and immediately adopt a whole new approach to life is absurd. Multicultural countries that appreciate the value of different cultures in society have a much more realistic chance of integrating newcomers into the established culture successfully than the melting pot idea. Integration in a multicultural policy country accepts the idea that new cultures will subtly change the entire culture of the country over time not simply be absorbed into the dominant way of life leaving it unchanged. After all, it is understood that cultures are dynamic not static and constantly change or die out. Adding new cultures into multicultural countries is a rejuvenation of the original founding culture not a takeover of it. Massive educational campaigns on the part of governments at all levels and by civic institutions are a necessary ingredient in preparing the way for the acceptance of such policies.

On a more programmatic level, fear of the other can be reduced though positive contact. Multicultural festivities and festivals open to the general public can be useful events to demonstrate aspects of new cultures that hopefully may reduce xenophobia on the part of the population. Recreational events of this type are useful policy tools in the quest for understanding and acceptance. Specific national policies can be generated and implemented to combat the inclination of migrants to carryover old wars and conflicts from the home country to the new one. School programs are a very valuable tool to address these issues. Black history months are an example of such a program regardless of how inadequate they are at present. Black history could be integrated into the entire curriculum as could the history of other cultural groups, particularly the indigenous cultures that were the founding peoples in the host land. If these types of programs were to be enacted, the curriculum should extol the positive characteristics of the culture and not portray them as victim. The airing of old grievances, while accurate, may not lead to lessening the fear and anger that perpetuates the xenophobia that lies at the root of the culture and immigration problem.

Migration is not going to lessen in the near-term but only grow until a new world order is created that reduces some of the causes that are perpetuating it. Perhaps the major causes are war, poverty, and climate change. Until they are addressed migration is likely to grow in size each year, so taking the aforementioned policy steps might go some distance in increasing the comfort level of the host peoples as well as providing a reassuring welcome to the new arrivals to their new land. Certainly, attempting to close down borders will only heighten the problem not fix it, nor will the xenophobic rhetoric spewed by the leaders of some of the host countries that experience immigration unrest the most.

Bibliography

Barber, B. 2001. *Jihad vs. McWorld: Terrorism's Challenge to Democracy*. New York: Ballentine Books.

Piketty, T. 2014. *Capital in the Twenty-First Century [Kobo version]*. Kobo.com.

United Nations. 2013. *A New Global Partnership: Eradicate Poverty and Transform Economies Through Sustainable Development*. United Nations Sustainable Development Knowledge Platform. http://www.un.org/sg/managment/pdfHLP_P2015_Report.pdf.

PART 2
Processes of change

7
RECENT CHANGE EXPERIMENTS

Recently, there has grown a worrying trend emerging in societies across the globe. Based on festering grievances by a growing number of people, a right-wing populist politics has grown in strength throughout North America and in many parts of Europe. This new movement is a result of many citizens feeling they have been left behind materially while a small group in society have benefitted greatly from economic globalization. Economic inequality, in conjunction with a growing perception of excessive immigration, is causing great anxiety in many countries across the world. The present inequality comes on the heels of the prolific development of a large middle class created at the end of World War II. Higher living standards through the transition of an agriculturally based economy to high-paying manufacturing jobs and the establishment of a social safety net policy are the hallmark of this reconstruction period. This newfound prosperity by a sizable proportion of society denoted the arrival of what became known as the middle class society. It appears that the reconstruction period has ended, and society is now entering a more turbulent period. A continual rise in living standards for the majority of western society is no longer assured. Social justice for many has deteriorated and we appear to have a society filled with anger and fear. Perhaps most importantly, the planet is reaping the consequences of decades of dumping CO_2 in the environment leading to global warming.

Liberal democracy does not appear to be fulfilling its promise to the middle and poorer classes. The economically marginalized no longer feel they have equal standing in their own country given the economic, demographic, and ethnographic changes taking place. Much of the established citizenry have become anxious about the recent influx of people who represent different cultural practices or who look different than them. It is clear that the demographic makeup of most western democracies is changing. The white European majority is now threatened by a growing coloured population and increased immigration from non-traditional countries is quickening that process. The population of many western countries find themselves alienated in their own land. Because of their social and economic deterioration, they have become angry at their governments who they blame for their declining situation.

As people become disenfranchised and marginalized in society, politics has become more pronounced and antagonistic, perhaps even violent. An unhealthy reaction to the faltering condition of the middle class has emerged as they attempt to address the new social, economic, and technological order that has had profound negative results on their station in society. A new right-wing populism and nationalism are on the rise and being promoted by leaders of some western countries, including the United States.

Nationalism 'takes the form of constructing links between one side each of the antinomies into fused categories, so that, for example, one might create the norm that adult White heterosexual males of particular ethnicities and religions are the only ones who would be considered 'true' nationals' (Wallerstein, 2004, p. 63). Nationalism promotes populist governments that have a penchant for isolationism. It is important to point out that historically nationalism has been a platform for the most horrendous practices of humanity. One only needs to be reminded of the Nazis in Germany and Mussolini's Italy to understand the potential consequences of over-zealous nationalism.

Populist movements have sprung up in many western countries. These movements embrace the idea of nationalism and isolationism, and this sentiment has infiltrated the politics of the United States and is being flirted with by many other countries throughout the world. The present motivation for populism is the reaction against globalization (universalism) of the economy which has been blamed for many of the ills felt by the marginalized population. Globalization it seems, has benefited the rich but not the average person. Populism is not just about falling behind economically, although that is a concern, but it is also about self-identity and one's station in the community. With the introduction of what is considered foreign values in society through globalization, anxiety among many of the long-term residents of affected countries has increased. Perhaps the most succinct way to summarize it is to suggest that fear of the other, anger toward leaders of the country, and feelings of being left behind economically and socially, are the bases for the rise in right-wing populism and nationalist sentiments.

Although society is faced with many worries and needs to embark on a major social transformation, the rejection of cooperative international action and a retreat to political and social isolation may not be the appropriate answer to these difficulties even though it may look good on first blush. In fact, such a movement may be antithetical to what is needed to positively advance the human condition. The populist movement in the United States, and in other developed countries, is an attempt to reverse course and retreat to a happier time when parents could be sure their children would do as well or better economically in their lifetime than they did in theirs. This does not appear to be the case any longer, so resentment and fear seem to be on the rise among those who are striving to remain in the middle class or join it. There is a heartfelt desire to retreat to the past as the slogans 'America First' or 'Make America Great Again', suggest. The populist notion of nationalism and isolationism is a powerful draw when things seem to be slipping away, but such a dangerous course deserves unpacking and critique. To begin this examination a review of two major western countries that are grappling with these issues can be instructive. Right-wing populism and nationalism in the United States and the anti-globalist and nationalist sentiments in Britain are two examples to be explored regarding the insights they may provide for the anxieties found in many western democracies today.

It is often suggested that to begin to solve a problem you first need to make it worse. US President Donald Trump may have begun to speak to the inadequacy of the world-system by making the present one even worse since his administration has taken power. The Trump phenomenon is the embodiment of the 'America First' and 'Make America Great Again' populist, nationalist, slogans. This ideology pits one country against all others. In this type of world there is no cooperative spirit to create a better world only competition that views international affairs as a zero-sum game with winners and losers. In fact, Trump often describes some people and other countries as losers.

Unlike other candidates in the 2016 US election, Trump tapped into the grievance sentiment that was rampant in many parts of the country, and I dare say, many parts of the developed world. Those who voted for Trump were clearly tired and distrustful of the standard political talk, generally about better days ahead, by politicians not paying enough attention to the feelings of the disenchanted. Trump supporters, and I dare say a large minority of the general population, are tired of hearing the well-worn rhetoric of 'a good life just around the corner' that permeates the rhetoric of orthodox candidates, who then do little or nothing to address the situation after being elected. Many people feel the promise of liberal democracy has betrayed them. Trump voters intuited that life for them is not on an upward trajectory but headed in the opposite direction in the globalized neo-liberal democratic world. They see their only path to be one of circling the wagons in an attempt to return to a bygone era. The tragedy of course, is that the imagery of a bygone era is just that, an image, and not something concrete to be retrieved.

A more sinister interpretation of this movement and its slogan 'Make America Great Again' is code for 'Make America White Again'. The election of Barack Obama, a black American, concerned many white US citizens to the point where they felt they were on the precipice of losing dominance in their own country. Many Americans have never accepted the African American as truly American and an equal citizen. The election of Barack Obama created a seismic shift in the political system and shook this sector of America to its core. To a certain extent the election of the Donald Trump government was a reaction to the Obama presidency. As a result, Trump, immediately upon election set out to dismantle as many of Obama's accomplishments as he could regardless of their merits. He signalled his intent to extinguish as much of the memory of the Obama presidency as possible. Much of the green revolution initiated by Obama was dismantled by Trump because it was Obama's legacy and not necessarily on its merits.

The cultural shift that confronted many Americans was compounded by an alteration in the economy for which they were not prepared, from mainly manufacturing to a high-tech service economy, including a heightened focus on finance. While these factors in themselves need not be debilitating, the forces that have sprung up around them are. Ideological gridlock that is motivated by the forces outlined herein suggests to some that the United States is now engaged in a cold civil war. Leaving aside the xenophobia and racism that appears rampant in many democracies today, much of the antagonism that exists in society is a struggle between those who believe society needs to embrace the new world with all its technologies and those who want to go back to an earlier era when the world was less complex. There appears to be little room at the present for negotiation.

The nationalist agenda has emerged and is clearly embedded in the Trump administration and pushing for what some pundits suggest is the United States moving toward a nationalist state and in the direction of authoritarianism. As Jeffko describes,

> Trump[']s America First message is essentially one of egoism and selfishness. It is anti-community. His message does not see the world as a global community of nations in the making, but a loose collection of nations motivated by economic and political self-interest striving for power and advantage over the others. Any form of cooperation for the sake of a greater communal good is suspect, be it NATO, NAFTA, the Trans-Pacific Partnership, the United Nations, or the Paris Agreement on Climate Change.
>
> *(Jeffko, 2018, p. 20)*

Given the emboldened actions of Trump following his impeachment acquittal, he appears to have taken on some characteristics of a despot. As J. S. McClelland summaries,

> [D]espotism is the rule of one man without the restraining force of laws – despots can make and break rules as they please, so despotism is government by caprice and whim. Any government, whatever its formal constitution, could act despotically if there were no institutional counterweights to moderate its force in practice.
>
> *(McClelland, 1996, p. 656)*

While there is a Constitution and democratic institutions in the United States, Trump tests the rules and is often quite successful at skirting them. He is able to intimidate the members of the Republican Party who now defend his actions unequivocally. The ability to engender fear is a feature of the despot and Trump has adopted this practice quite well. Additionally, he has been able to appoint far-right leaning members to the Supreme Court where they are now in the majority and tuned into many of his ideas about how the presidency and America should work. He has also purged his cabinet of secretaries who showed allegiance to the Constitution for members who demonstrate allegiance to him, including the Attorney General. Additionally, and as mentioned previously, Trump has vilified the press designating them as 'fake news' and the 'enemy of the people'. These are all signs of despotic behaviour.

What the US version has in common with authoritarian governments from history is xenophobia, ethnonationalism, a deepening of the separation between the classes, and an attempt to control the press to create a narrative that suits their purpose whether it has any relation to the truth or not. Like other such nationalist regimes, the Trump administration provides the public with versions of reality that may not correspond to observable reality. Unlike other less democratic countries that are not embedded in the rule of law, the United States must operate inside a Constitution that limits just how far an administration can go, unless they alter that document, and there is no sign of that happening yet. That said, the Trump administration has clearly usurped congressional power as evidenced by his refusal to cooperate with the oversight function of Democratic-controlled House of Representatives through denying members of his administration from testifying or in providing documents to many of their committees even when subpoenaed.

Trump seems to revere the authoritarian leaders of other countries. The king of Saudi Arabia; Putin, the President of Russia; Kim Jong-un, the leader of North Korea; and Xi Jinping, the leader of China, all authoritarian leaders, are the kind of people to which he seems attracted. Early in his political career, he suggested that Russian president Putin was a stronger leader than Obama. At the same time, he has developed an antagonistic relationship with most of the leaders of the allied countries. What this move toward authoritarianism seems to indicate is that a portion of the US population is seeking nothing short of revolutionary change, and this is what Trump represents to them. The two contesting groups in this struggle are the alt-right who desire a populist and nationalist government and the more moderate citizenry who understand that their future lies in a globalized world, but an altered globalization directed from the bottom, not the top, as is now the case. Whether or not the Trump version of social change leads to the most appropriate transformation and moves the country toward a greater prosperity, equality, and social cohesion is highly unlikely, but only time will tell. I suspect that Trump was the only candidate for president who appeared to be anti-establishment in the minds of the voters, and that provided the key to his success. Trump's rhetoric was good enough for those losing ground to log their deep dissatisfaction with the direction in which the country was headed. Many of the Trump supporters remarked that he was someone who appeared to be like them. Let there be no mistake, many of the social ills identified by the Trump supporters are valid and need attention. The question remains, will the Trump version of government accomplish what his supporters desire?

The Trump crusade campaigned on retrieving manufacturing jobs that had left the country and gone to the developing world. What was clearly misunderstood was that many of the jobs that left decades ago no longer exist but have been replaced by technology; or, if those jobs still exist, they are being paid at such a low level that no worker in the developed world could afford to take them up if, in fact, they were to return to the United States. Also to be considered, the present administration is generally anti-government and a pro-private economy administration. Certainly, there is nothing unusual about that; western economies have been increasing their confidence in the private sector for decades, maybe to the extreme. Trump has placed cabinet secretaries in posts that have vowed to deconstruct the very departments they are now overseeing. These policies of deconstructing government bureaucracy cannot be inspiring to long-term employees who are dedicated and have spent their careers attempting to serve the American public in a variety of ways. We see just how understaffed and underequipped the frontline workers were during the COVID-19 pandemic. The Trump administration has not filled several of the senior positions in many of these departments, particularly the State Department, thereby lessening their effectiveness in world affairs. All that taken into consideration, history suggests that world supremacy is not everlasting but has a time horizon. Perhaps the best example of this phenomenon is the rise of the United States and decline of the British Empire after the end of World War II. The present attempt of the United States to hold on to their dominant position in the world may be inconsistent with history.

What is disheartening but instrumental to this discussion of social retreat and reaction to the stagnation and decline of the working middle class in society is the fact that the Trump campaign ran on bringing coal and other forms of highly emitting CO_2 fuels back into production on the pretence of creating jobs. This commitment was a concrete demonstration by Trump of the abstract idea of Making America Great Again. This policy led to the denouncement of global warming as a human-made phenomenon. After his victory for president in 2016, President Trump withdrew the United States from the Paris Climate Accord. He has also eliminated many of the regulations under the auspices of the EPA, which is the department responsible for combating environmental deterioration. Trump has also named many cabinet members to important positions in his government, including the secretary responsible for energy, who have been, or still are, climate change deniers. Many of those holding influential positions in the Trump administration were also lobbyists for the fossil-fuel industry before taking their present government positions. Economics and job creation hold sway over any notion of combating climate change in the present administration. In fact, I'm not sure climate change is on the radar at all beyond occasional gratuitous remarks and there haven't been many of those lately either.

While there are numerous elements that characterize the populist notion of retreating to earlier times, I will only introduce a few of the most notable ones here. The first is xenophobia, some would say pure racism. The issue of xenophobia and the backlash toward immigrants particularly, is largely focused on the Muslim population but also on people migrating from Central or South American, countries that are in social and political turmoil or under economic stress. This is one of the hallmark characteristics of an ethno-nationalistic and isolationist country. In Trump's regime, the argument for banning Muslims from entering the United States attaches itself to the rhetoric of keeping the homeland safe from terrorism. This policy focuses on particular countries from the Middle East and North Africa, none of which had anything to do with the attack on the Twin Towers in 2001, nor have they produced a terrorist act on US soil in the past. Other North African and Middle Eastern countries whose citizens have perpetrated acts of violence against the United States are not on the ban list. How do we explain this? Some would argue that the countries whose citizens have perpetrated terrorist acts against the United States but not on the ban list are those countries that supply oil or possess other commodities that the United States needs and wants. Trading relationships appear to trump (no pun intended) terrorism as a rationale for exclusion from the ban. Fostering military alliances is also an important factor in a region that is often hostile to the United States. Some of the countries thought to be allies of the United States, such as the Kingdom of Saudi Arabia, have terrible human rights records to which the United States has traditionally objected, in addition to their alleged funding and sponsorship of terrorism throughout the world, but they have escaped the ban nonetheless.

Xenophobic and anti-immigration policy is frequently considered to be nothing more than the creation of an external enemy that can be used by government to refocus their inadequacies onto an alien country rather than coming to grips with internal problems. This is not a new strategy the Soviet Union served this function for the United States for decades throughout the cold war period from the middle of the 1940s to the early 1990s.

Recession is a term that is most often, if not exclusively, used to describe two consecutive quarters of negative growth in GDP. Ian Bremmer (2018), an American political scientist, has recently used the term 'political recession' to describe the state of geopolitical affairs. Certainly, the morass created by Brexit and some of the other conflicts in the European Union fits into that category, but I think the term best applies to the confrontation between China and the United States. Although this conflict is most often discussed in economic terms, and it is certainly fought through economic measures like tariffs and sanctions, what is occurring on the world stage is more than an economic dispute. It is, in fact, a fight for world geopolitical hegemony. China is making a bid to overtake the United States not only in economic standing but also in terms of world power and hegemonic domination. Because of this new cold war, many 'normal' parts of daily life are in jeopardy or at least made uncertain for many people. Employment and identity based on cultural allegiances is under siege. The world balance of power is being reframed. China is moving away from being strictly an exporting economy toward becoming a consumer society as well. It is also extending its military strength and becoming a regional power by replacing the influence of the United States in the South China Sea and other strategic parts of the globe. China is constructing military islands on atolls and reefs in the South China Sea which threatens other Asian countries like Vietnam and the Philippines. It continues to harass foreign military ships and planes as if it has already claimed ownership of this space. This power shift from West to East is occurring as the United States withdraws its military and political power in the region (Kaplan, 2014). Although it is still too early to determine who will be the victor in this struggle for world domination and hegemony, there are several factors that provide China with a leg up. The leadership in China is stronger and more experienced in geopolitics and world affairs than the present government of the United States. Because of the political structure of China, it has greater capacity to absorb pain than the United States. The Chinese leadership has a longer time horizon than the US administration to be in power, which gives them more time to wait out their adversary. Certainly, the citizens of China show more discipline than the US population because of their form of autocratic government. The size of the US debt, which is growing by leaps and bounds, will surely put the United States at a clear disadvantage as this debt grows even larger. Debt is also a factor for China, but it may not be as much a concern as it is for the United States. Given the massive increase in debt caused by the COVID-19 pandemic for both countries, the debt picture may no longer be as clear as it was prior to the detection of the virus. How this massive influx of debt will affect both countries is not likely to be clear for some time, but the United States may have bitten off more than it can chew if this political and economic conflict continues in the long-term.

The Trump regime has also been particularly good at demonizing the media. The administration has attempted, with some degree of success, to designate the media as 'fake news' and as 'the enemy of the people'. While politicians have traditionally had an antagonistic relationship with the press, the populist regime takes this antipathy to the extreme. Only the news outlets that support the regime are considered legitimate. The others are not only seen to be antagonistic but in cooperation with those who would depose the administration. To counteract the 'fake news media', Trump has adopted the FOX News network as his outlet for broadcasting his version of reality. The so-called reporters on the Trump channel create a highly partisan image of the president and his actions and the outcomes of those activities. Authoritarian leaders down through history have created their own narrative of events and revisionist history to forward their agenda and maintain their legitimacy.

The hallmarks of the type of politics embodied by the populist Trump regime, or as some would claim, a social movement rather than a political program, are movement toward isolationism and nationalism; a focus on the past when the mass population was generally ascending in social position; the denial of the existence of any phenomenon like climate change that threatens their major focus on growing the economy; xenophobic and anti-immigrant; the creation of external enemies on which to focus the population's attention rather than the inconsistencies that arise at home; and the creation of an internal enemy such as the media or a corrupt judicial system to blame when things don't go the way they are promised.

Early on in Trump's tenure as president, Robert Mueller was appointed to conduct a US Department of Justice investigation into the possibility of illegal Russian involvement in the 2016 elections (Mueller, 2019). Six people who worked on the

Trump campaign in one form or another were indicted for wrongdoing or pleaded guilty as a result of that report. A number of those convictions were a result of individuals lying to Congress or to Mueller's investigation team. A number of Russians were also indicted but were never extradited to the United States for trial. Roger Stone, a long-time friend and business associate of Trump, was convicted and sentenced to prison for lying to Congress about his association with Wikileaks for the nefarious purpose of publishing hacked Democratic Committee emails. He has subsequently had his sentence commuted by the president thereby escaping prison time. Trump himself was not indicted for the activities of collusion or obstruction of justice as asserted by his adversaries but was identified in the Mueller Report for a number of misdeeds that were later part of the impeachment trial in the House of Representatives. Trump was impeached by the House of Representatives for allegedly asking a foreign government to undertake the investigation of a political rival to advance his own political ends but eventually exonerated by the Senate. The impeachment of Donald Trump was a heated partisan process.

The authoritarian regime takes no responsibility for things that go wrong. All problems are said to be the fault of someone else. The demonization of particular groups inside and outside the country refocuses the plight of the disenfranchised on a non-existent enemy. This projection of bad behaviour onto others occurred in conjunction with the rejection of traditional moral values by the country's leadership. This became clear with the release of a video in which Trump, without impunity, was overheard making what one would think to be unacceptable sexual remarks, degrading women, and giving license to sexual assault. The group of voters who are attracted to Trump have also rejected any thought that the longstanding plight of minorities, especially African Americans, should receive special attention to redress the problems they perpetually encounter particularly from the law and judicial system. The attack on political correctness has allowed the darker side of bygone days to remerge. Schools and other social institutions are reporting greater hate speech and hate crimes than during earlier administrations. Self-professed white nationalists, including the Ku Klux Klan, have praised Trump's agenda and rhetoric.

In authoritarian governments, the creation of a strongman to head the regime is a prominent feature. In addition to Trump emulating this feature in the United States, there are a growing number of right-wing populists and authoritarian leaning political parties across the globe engaged in that practice. The Netherlands fought off such a leader in 2016 as did France in the form of Marine Le Pen that same year. This is also a growing phenomenon in some eastern European countries as well. What most of these movements have in common is an outright rejection of what they have termed the 'Deep State'. This too has been a focus of the US administration. Although this term has not been defined officially, Wikipedia describes it as a coordinated effort by career government employees and others to influence state policy without regard for democratically elected leadership. Consequently, the bureaucracy is deeply mistrusted by the alt-right and in some cases ignored or terminated. The Deep State in the United States is just another one of Trump's boogeyman. This mistrust was demonstrated by Trump's attitude and behaviour toward the intelligence community shortly after he took office, and his rejection of this institution continues today. Trump has taken the word of his international political adversaries, particularly Vladimir Putin, over his intelligence agencies on a number of occasions. There is clearly a mistrust of the bureaucracy. A purge of the senior bureaucrats is often a central feature of the populist government. Certainly, Trump's initial White House appointees, with few exceptions, had no government or military service and came straight from the private sector. What is particularly curious in the Trump cabinet is the number of billionaires or multimillionaires with whom he has surrounded himself, although they consider themselves a populist government that claims they intend to rectify the plight of the middle class and those marginalized in society. Their policies would suggest otherwise, however. They have given a huge tax cut to the largest corporations, and most of it, estimated at 80%, will go directly to the wealthy, not the middle class or poorest in society. A major slogan of the Trump campaign often chanted by his followers at his political rallies was, 'drain the swamp'. While the populist rhetoric is in full view, the behaviour of the government would indicate something quite different. Early in the Trump administration, many programs focusing on the poor and the environment were designated for budgetary reduction in favour of increasing military expenditures considerably. Most notably the administration engineered a huge tax cut, most of which unapologetically benefited the rich. It would appear, to at least this point in time, that what we see is a wolf in sheep's clothing. Until now there has been great distance in what Trump campaigned on and what he has done since being in office. When he failed to initiate many of his promises, which were actually few in number, he has blamed the Democrats for those failures. Despite these glaring inadequacies, he maintains his base of supporters. It can be argued that style and rhetoric make those who are aggrieved feel better psychologically, which appears to be more important to them than policy or programs that may ameliorate their social and economic condition.

    The forgoing is a description of the Trump government and is meant to describe the features of a right-wing populist and nationalist political philosophy that resonates with many voters who possess very legitimate concerns about their economic decline and deteriorating station in life. There appears to be a feeling of competition between the people and the government. In the view of Trump supporters, the government bureaucracy has grown at the expense of the people. What is of interest is that Trump supporters don't see the predatory capitalist corporation as having a role in their plight. They recognized that they have lost ground over their lifespan even though the rhetoric of the American Dream suggests that their economic and social status should be on the rise. The social system has failed them, plain and simple. What they had bought into, liberal democracy, did not deliver on its promise. Trump promised something quite different than the platforms of most politicians. Despite those promises, the situation continues to deteriorate. Hopefully, the Trump phenomenon is a momentary sidetrack and the United States, and other countries that are in the midst of a similar struggle, will determine that right-wing populist regimes with agendas like the one outlined herein simply mires them deeper into despair and does nothing to rectify their plight even though the rhetoric may make them feel good psychologically for a short time. Good feelings often come from blaming others for your plight. Although in some cases there may be others who are responsible for degrading conditions, it is usually not other marginalized people who hold the key to the downward spiral. This social decline is not necessarily a new phenomenon, although it may be unique for a country such as the United States. This scenario has played itself out around the world many times before, including in most countries of South America and other developing regions across the globe.

What may be strikingly unique about the situation in the United States is the apparent abdication of their leadership of the free world. It must be remembered that the United States has had an isolationist history. Isolationism was very strong, leading up to World War II. Although eventually the United States was a large factor in bringing the war to a conclusion, it entered the war late. There was great resistance on the part of the public to enter a war on the European continent.

Part of this discussion will need to examine the speculations by some that the world's leadership is moving from the United States to elsewhere. Where this 'elsewhere' lies is clearly not determined yet, but there is speculation regarding China and Europe and their potential ability and willingness to pick up the leadership mantel. Whether either of these political areas of the world have the wherewithal or political will to engage in such an overhaul of world leadership is yet to be determined. The world may be left to drift leaderless until a new world-system emerges. The abdication of global leadership by the United States is another attempt at rejecting internationalism and the idea that economies and countries are intertwined, and the future is global not just local, is rejected. This retrenchment can be seen by the use of the word 'again' in the phrase 'Make America Great Again'. Clearly there is a feeling that something has been lost and needs to be regained not by joining and leading the world in new directions but in trying to take back what was thought to have been given away or stolen. This sentiment is also contained in the rejection of the notion of global warming by calling it a hoax perpetrated by some other far-off government – in Trump's mind, China.

It appears that the traditional right and left political spectrum is breaking down. The rise of the alt-right (alternative right) in the United States and other developed world countries deserves some examination for its influence on the political scene. It is certain that the alt-right has had a significant influence on the politics of the United States over the last decade. The outcome of this movement has been the creation and rise of the Tea Party as a powerful faction in the conservative Republican political party. Some would argue it has taken over the Republican Party. This takeover has essentially caused gridlock in government. There is no longer room for compromise. The alt-right in the form of the Tea Party was instrumental in electing Donald Trump to the position of President of the United States. The government has taken into the administration members of the alt-right as senior

120 Processes of change

advisors to the president, who are dedicated to the deconstruction of the Deep State, as they term it. How long the takeover of the democratic government by the alt-right and right-wing populism will last is unknown, but while it lasts, it will have dramatic effect on the world-system.

A major question to be addressed is the long-term effects on the American presidency, perpetrated by the autocratic leadership style of Donald Trump. It can be argued that Trump's stamp on the US presidency will be seen as wide ranging and long lasting. During Trump's tenure, the presidency has increased the power of the executive branch of government, usurping the powers of the other two branches. In a sense, he has become an Imperial President. He continually tests the constitution; unilaterally withdraws from international agreements such as the Iranian nuclear deal, the Trans-Pacific Partnership, and the Paris Climate Accord; appoints cabinet members who are loyal to him, not the Constitution; uses his power to punish members of Congress who dare to oppose him; pardons friends and white-collar criminals; regularly interferes in the affairs of the Department of Justice; continually criticizes judges and the judicial system; has weakened NATO and long-standing relationships with America's allies; supports other dictators throughout the world, particularly Russia and North Korea and criticizes allies; has tried to gain political advantage through the use of foreign governments; declares the media as the enemy of the people; has supported the white nationalist cause through his failure to condemn their practices; and other such undemocratic activities. The issues raised herein have changed the office of the President of the United States dramatically and permanently. Once these highly contentious practices have been employed without pushback or consequence, they will be easy to implement again, perhaps even made permanent. It is like a rubber band – once you stretch it to its limit, it will never return to its original form. Even though the tenure of Donald Trump in the White House is terminal, he has changed the power structure of the presidency, and perhaps the idea of liberal democracy itself.

A second recent example of the populist and nationalist sentiment sweeping across the western world can be seen in the decision of the British people to succeed from the European Union. The UKIP, a populist movement, tapped into, or indeed generated, a backlash against the integration of Britain with the rest of Europe. Many of the western European countries came together initially to create a political entity which became known as the European Union. The European Union is an outgrowth of the European Economic Community (EEC), formed by the six countries in 1958. The EEC itself was an outgrowth of the European Coal and Steel Community (ECSC) which was formed in 1951. The European communities (EC) became official under a treaty that was signed in Maastricht in 1993. The latest version of this integration came in 2009. It created the European Union as an overall legal unit containing a bill of rights and governance structure along with defined decision making authorities. While I have not presented in any detail the various permutations this arrangement went through over the years, it is important to note that integration of the European countries started in 1948 after World War II with the Brussels Treaty. Essentially, this movement was designed to bring the countries of Western Europe together through cooperation to avoid, yet again, another war on the continent, given that it had just gone through two major conflicts. Since 1948, the countries of Western Europe became increasingly integrated through sharing of some major common policies, particularly around trade and economics, migration and immigration, human rights, and a common currency. Some of the eastern European countries joined the European Union after the dissolution of the Soviet Union in 1991. Britain was a late signatory to this union and did not participate in all its features. Perhaps most notable was Britain's non-acceptance of the Euro as its currency, as it preferred to maintain the Pound Sterling. Even though all signatory countries elected members to the EU parliament, many British people were always somewhat sceptical of the European Union and felt that too much policymaking was given over to Brussels, the site of the EU parliament.

The unease of membership in the European Union among some British citizens caused David Cameron, then British Prime Minister, to promise to undertake a referendum before it would sign the final documents and become a full member of the European Union in 2016. Most political pundits believed, as did Cameron, that the referendum to join the European Union fully would receive overwhelming support by the British people. Little did he know that he was in for a tough fight and an eventual surprising loss.

I think it is fair to say that until this period in history, the UKIP had always been a marginal political party in Great Britain. It was headed by Nigel Farage, a staunch antagonist to the idea of the European Union even though he was elected to represent Britain in Brussels. He was also a confidant of Donald Trump and is a staunch supporter of the nationalist cause. Farage and his party took up the 'leave the EU' campaign during the referendum in the fight against Britain becoming a full-fledged member of the Union. The term Brexit (British Exit) was coined as the rallying slogan in the fight against the forces in favour of fully joining the European Union.

What were the characteristics of those who voted to stay in the European Union and who voted to exit? Many rural and non-urban residents were in favour of leaving the European Union. 'Leave' voters were also older, poorer, less educated, and far more likely to think the country was getting worse off than those who thought it was doing well and who saw the future of Britain solidly attached to Europe. People who lacked opportunity and marginalized in society were among those who supported the leave side. Urban centres, particularly London, chose to remain in the European Union as did Ireland and Scotland. The leave voters were clearly those who felt sidelined in their own country and who felt they were being taken over by a foreign entity, and Britain's integration into continental Europe was not seen to be working for them.

As the campaign took root and played out over time, it became apparent that the fight was not simply over such issues as whether membership in the Union was economically advantageous or not but took on more sinister elements. Xenophobia and immigration became a major part of the debate. Many British people felt they had lost their jobs to immigrants and were resentful of that fact, whether it was true or not. Although Muslim immigrants were a target of this xenophobia, so were other nationalities, particularly from eastern Europe. The type of xenophobia that Latinos and Muslims were experiencing in the United States, as described earlier, was being exhibited in Britain as well. While a slogan, such as 'Make America Great Again', was not verbalized in Britain, the same sentiments were being expressed in British terms. People yearned for the 'good old days' when employment of the young entering the workforce was assured, and the people who surrounded you every day looked like you, and the future looked bright or at least brighter than it did at that moment. As in the United States, young Muslim people were leaving Britain in large numbers to travel to the Middle East to fight with ISIS in order to create an Islamic caliphate. Some radicals inspired by ISIS were planning and carrying out terrorist attacks in British cities. That also played into the rise of xenophobia.

The separation of the socio-economic classes and religious divisions became more pronounced than in earlier times. So, while Britain did not elect a populist leader who promised to retreat from the world and focus solely on the homeland as was the case in the United States, similar sentiments were expressed during the Brexit campaign, and a good portion of the population voted for isolationist, nationalist, and populist policies embraced by the leaders of the Brexit movement. Needless to say, Brexit carried the vote. Boris Johnson, an ardent Brexit supporter and eventual replacement of Theresa May as Prime Minister of Great Britain, leans farther toward the populist ideals than his predecessor although he doesn't seem to have the hard edge that Trump demonstrates.

Although Brexit is now a fact, there is yet no agreement establishing the rules under which Britain's trade and other relations with the European Union and other countries throughout the world will be conducted. These agreements will likely be harder to accomplish than the initial withdrawal from the European Union and may take some time to accomplish. In all likelihood, there will be winners and losers in these agreements as there are in all negotiations. Whether or not the British people will be better off after the conclusion of these agreements is still unknown, but most observers feel that Britain will endure hard times for a while. Certainly, the Governor of the Bank of England during the Brexit debate warned the country of an uncertain future if the Brexiters won the vote, which they did. Isolationism in the past has not aided countries in their desire to advance economically or socially.

The fundamental issues centred around economic globalization and the perceived lack of benefit of this arrangement for many, and the immigration issue, were the root causes of Britain leaving the European Union. There was also great concern that Britain was giving up too much power to Brussels and not getting enough in return, although the latter concern may not be true. Again, the perception, whether true or not, is that only corporations and the wealthy have benefited from international treaties like the European Union in Europe or NAFTA in North America, and the immigrant is destroying the social fabric of the country. Treaties that go beyond economic agreements such as the Paris Climate Accord are also seen to favour those with wealth and not those who lack enough resources to carry out their daily activities. This is not to suggest that members of the wealthy class are in favour of environmental regulation; they surely aren't. Their objection is a bit different, however. They are solely interested in the economic imperative and are blind to the environmental damage their activities cause, or they are sufficiently narcissistic to simply not care.

I have laid out two recent examples of countries under political and social distress. Both are turning to a more right-wing populist and nationalist form of government. Populism, isolationism, nationalism, and xenophobia are often driven by the lack of social cohesion where all members of society feel they are being treated fairly and have a stake in the social system. Many of those characterized as marginalized in both cases presented earlier do not feel social cohesion with the majority of their society and often feel they have been left behind or left out completely. As a result, they are angry and getting angrier as time goes on without recognition, let alone resolution of their plight. What did we learn from the US 2016 presidential election and the Brexit vote? Are there fundamental principles that come out of these cases? Are there some common themes? These are the questions that need to be addressed as we analyze the inadequacies of the world-system under which we live. The United States and Brexit examples may provide some clues as to the characteristics and themes that may need to be integrated in any system that facilitates large changes to the world's social organization structure. Before laying out those characteristics, it is important to clearly articulate their underlying elements.

Humans have unique features that separate them from other primates; culture, which contains sophisticated *symbolic communication*, including language, is certainly one of those characteristics. Without going into the nuances of human communication unnecessarily, it is important to note that language requires several features to make it useful and purposeful in the process of communication. Perhaps most importantly, language allows humans to engage in what psychologists described as '*the theory of mind*' (ToM), the ability of one person to understand the meaning and intent of another person's dialogue and, in turn, attributes the corresponding ability to the other person in the discussion. It may not only be what is said but also the ability of the other person to relate emotionally to the speaker's speech and vice versa that allows for full understanding of the communication. This process is unique to humans. The Trump and Brexit phenomena have introduced xenophobic and racist language at a heightened level into the political debate. Language allows a person to clarify in their own mind what they think and feel about a subject or condition that confronts them. Can you imagine what the process of thinking would be like if you couldn't verbalize your thoughts in your own mind? Without language, it may not be possible to think in the abstract at all. Some of us even verbalize our thoughts openly and out loud, a practice that is often viewed unnecessarily with alarm. Often, what is verbalized by the other person in the dialogue is what you are thinking consciously or unconsciously. So, language allows us to conceptualize and reflect on our feelings as we strive to make decisions about everyday occurrences. Without language, this state would not be possible. To hear other people voice our own thoughts reinforces what we are thinking and, perhaps, emboldens us to state them forcefully out load. The alt-right in many of the democracies of the West have capitalized on emotional language that resonates with many in society, particularly those who feel betrayed by the ruling class. When someone else verbalizes what you are thinking in the abstract, it gives you license to do the same. In the recent past, anti-immigrant speech was less acceptable than it is now. Nationalism and populism have given license to speech that borders on hate, and at times falls over that line. We see examples of that type of behaviour in both cases outlined herein.

In both the 2016 US election and the Brexit examples presented earlier in this chapter, the ability to conceptualize the condition that many citizens find themselves faced with was a necessary ingredient for creating individual willingness to change. Not only were many citizens of the United States and Britain willing to change the situation in which they found themselves, but sufficiently moved to demand participation in that change. Theory of mind accommodates both rational and emotional intelligence, and, while no one can say for sure, other species do not engage in ToM or rational and emotional thought. In many ways, Farage and Trump verbalized and hence legitimized what many of their followers were thinking in the abstract. That is part of their appeal. The language engaged in by both of them appeals to the darker side of human nature.

Although we all like to think that we act exclusively on rational thought, emotional intelligence plays a key role in the decision making process. In both examples cited earlier, fear and anger, both emotional components of the process, played a key part in the election of Donald Trump and for Britain to vote yes for Brexit. In the United States, many citizens were angry at how their situation in society had deteriorated. Those who supported Trump were people who believed that he talked like them and not like other politicians or elites in society. Things will need to be much worse before they abandon Trump, no matter what policies are enacted or not by his administration. It could be that the COVID-19 pandemic will give voters a chance to reflect on where their country is headed. In recent history, emotional intelligence has dominated their thought process because of their deep-seated grievances. They will remain with him as long as he continues to speak their language. The Brexiters also spoke to the emotional side of the common person. This carried enormous weight with those at the socio-economic bottom, or at least those losing economic ground. This is not a right-left split but an open (those who see globalization as a wind at the back of progress) or closed (those who see globalization as what has cost them their jobs and economic deterioration) phenomenon.

Despite where the Trump voters are located on the socio-economic scale, they don't seem to resent billionaires or the inequality that exists. One can only speculate that US democracy has been completely taken over by celebrity culture. Many, if not all, of the billionaires who populate Trump's cabinet have made their fortunes on the backs of those at the low end of the socio-economic ladder, precisely those who support Trump. It may also be that those at the low end of the economic ladder have faith in the 'American Dream' and still hold up these new-age aristocrats as individuals to emulate. As suggested earlier, it is the immigrant and globalization that are seen as the cause of America's problem and not the billionaires who have exploited the system over the years and are the key players in advancing the economic globalization project at the expense of supporting human rights and social justice.

The Brexit vote demonstrates that there was great fear in the form of xenophobia rampant in many regions, mostly non-urban, in Britain. No doubt xenophobia played a huge role in the US presidential decision making process as it did in Britain's decision to leave the European Union. Most likely, fear and anger were also deeply involved in the thought process in each case and interacted in a highly complex fashion to produce thoughts of populism and nationalism in the minds of many voters. As initially attractive as populism may seem as a way of combating one's declining station in life, it will not provide a sustainable remedy over the long-term. The social and environmental issues of poverty, the unequal distribution of wealth, and the international concern that has sprung up over climate change and global warming will not be addressed by a rise in populism and nationalism. Nationalism tries to shut itself off from what is going on in the rest of the world. It is a reaction to international institutions that have been established to mediate issues of global concern but appear to diminish the role of sovereign governments. These institutions are now seen to work against the worker in favour of those with wealth. What does not seem to be occurring today is any analysis or rational thought on these issues. For a good number of voters, emotional thought has taken over completely.

What is crucial to understand is that the elements of emotional intelligence and fear and anger are major players in the decision making process even though we humans like to think we act rationally when confronted with life-and-death situations. We can now partially understand why incongruent notions such as 'alternative facts' emerge to support one's emotionally arrived at decisions, rather than using reality to construct more rationally based decisions. The problem with employing emotional intelligence in decision making exclusively is that it often leads to false analysis, conclusions, and inappropriate solution selection that do not turn out to address the undesirable condition satisfactorily. In fact, it may even acerbate the problem. It may turn out to be that the outcomes of the 2016 US presidential election and Brexit fit that later category. Although those who voted for Trump and Brexit may feel good emotionally because they stood the status quo on its head and were contrarians, they may be very disappointed when their lives do not improve as expected but continue to deteriorate because of their investment in a wrong set of solutions. In fact, the policies that are being applied today may not even attempt to address the issues and problems of those who have been marginalized in society. The problems may have been improperly diagnosed. It often appears that xenophobia, nationalism, and populism are strategies looking for a problem to fix without clear understanding of what the end goal should be.

The emotional intelligence approach to decision making that embraces fear and anger as its driving force does not stand a great chance of producing satisfactory solutions in most cases as both the election of Donald Trump in the United States and Brexit would suggest. It should be stated, however, that there is a role for emotional intelligence in the process if used appropriately. Emotional intelligence plays a key role in alerting society that something has gone wrong in the social or environmental system. Certainly, this is what the Trump and Brexit supporters are signalling. Additionally, emotional intelligence may provide the motivation for addressing the inadequacies of society and in developing a social movement to attack them. Once that condition has been obtained, rational intelligence should be employed to analyze and investigate the problem and then, most importantly, select the correct set of solutions and design the right plan to resolve those issues. Emotional intelligence will also help inform the decision maker if the right decision has been made. It appears that there is an integration between rational and emotional intelligence when it comes to decision making. To not move between emotional and rational intelligence in decision making, as outlined earlier, is to design a program from a singular point of view which could cloud the reality of the conditions. However, the rational part of the equation is often missed in the decision making process. Humans may have a disposition to use emotional intelligence as their default mode in decision making because of evolutionary history. There was a time during humanity's early development when emotional intelligence needed to be dominant in decision making. It was advantageous to make split-second decisions based on emotions, such as fear or anger in order to avoid being eaten. Today's complexity demands an altogether different type of decision making strategy that necessarily involves rational thought. It may be that the emotions among Trump voters regarding the state of their lives overrode rational thought, impairing the selection of the alternative that would truly address their plight. In fact, there may not have been an attractive alternative to Trump that they considered adequate to address their situation. And I dare say, those who voted for Donald Trump will not see a significant material jump in their lives and the enthusiasm that carried him to office should soon wear off. Being a contrarian for the sake of being a contrarian or destructive force will not produce the needed changes that are at the heart of the problem in either the Brexit or the US case.

What the Trump and Brexit examples suggest is that the global human social system is engaged in a fundamental transition. This turning point is motivated by converging forces that are changing the system that was established by the early western democracies. The examples earlier suggest that even advanced democracies can slide into quasi dictatorships. While Britain does not show advanced signs of this slide, the United States has elected a president who has surrounded himself with advisors who are subservient to authoritarian tendencies. What is interesting to note is that Nigel Farage showed up on the campaign trail with Donald Trump during the 2016 US presidential election campaign, suggesting they were of like minds. Both heaped praise on one other and their jointly held ideology.

While it may be accurate to suggest that neither of these countries have instituted a strict dictatorship, they have certainly moved closer to plutocracy over the years. A plutocracy is defined as government by the wealthy. Based on this simple definition, it can be argued that the United States has gone the distance and has fully become a plutocracy. The ruling class has enacted policies and legislation that benefits them but not the middle and lower economic classes. The same cannot be said about Britain yet, at least not to the same intensity as in the United States. Britain's parliamentary system may not be as amenable as the US presidential system to that type of pressure. That said, British politics was born out of an aristocratic tradition.

Trump's cabinet is filled with billionaires, perhaps more billionaires than millionaires. He argued in a recent speech that he doesn't want poor people in his cabinet, but he particularly wants rich people in economic portfolios because they have done it for themselves in the past and can employ these qualities for the nation. It does not appear that those skills, successful in the private sector, are necessarily transferable to the public sphere. And, although the wealthy may understand what policies and measures would benefit their class, they may not understand fully the plight and lived circumstances of the poor and middle classes and what is needed to improve their lives. The US citizenry has long admired the super-rich, so initiating a plutocracy does not seem to bother the general public.

Even in a celebrity culture that admires the super-rich, the state of inequality that exists today is having a detrimental effect on social cohesion in the United States and other western democracies. We see spontaneous development of such protest groups as the Occupy movement and BLM and their derivatives in other parts of the developed world. And, while the recent version was short-lived, the motivation for its creation lives on. There is no doubt that this type of sentiment will rise again as long as there continues to be an increase in the wealth of the 1% and diminishing resources for the other 99% of society. Social solidarity has broken down across the democracies. It is arguable that the United States and maybe Britain as well as other western democracies are fast becoming failed democracies or, at least, in the process of reinventing themselves. Only time will tell which path each selects. Although there may still be some semblance of right and left in the political spectrum, the landscape may be changing to resemble groupings of like people. Those left behind in society appear to be seeking a strongman to take control and provide psychological, physical, and economic security in what appears to them to be a failed democracy. Those on the more progressive side are developing policy proposals (the Green New Deal) that are designed to create a more egalitarian society that deals with the social and environmental issues that are stressing the globe. There is now an alternative to Trumpism in the public forum that could move society forward rather than stagnating in the present or returning to some version of yesterday.

Extreme partisanship is now the new normal in most liberal democracies. Although partisanship has been present at most points in political history, it has now come to the point where gridlock reigns supreme, making progress on any issue difficult or maybe impossible. The gap between the political parties seems so wide that any form of compromise seems unlikely. Essentially, the gridlock with which our political system is afflicted reflects the condition of larger society. Citizens in many of the western democracies have never been so divided. When one branch of government is controlled by one party and the other branch controlled by a different party, gridlock is understandable. Today, we have legislative gridlock when the entire government is controlled by a single party given the factions within political parties today. Regions throughout the world also suffer from political gridlock, making peace-making in war-torn regions extremely difficult. The traditional political parties are now in turmoil and about to break up into other nuanced entities. Party politics is being replaced by tribal identities that have singular ideas and uncompromising agendas. Take, for example, the Republican Party in the United States. It now contains at least two factions one called the Tea Party and the other known as the Tuesday Group. Each of these groups encompasses a completely different philosophy of government, neither one compatible with the other. The Democratic Party is also showing signs of breaking apart philosophically, albeit not as fractionated as the Republicans. Although perhaps not as contentious, there are several streams in the party that are vying to take charge, and it is possible that the Democratic Party will also fracture into several competing groups. The Conservative Party in Britain also showed signs of stress fractures when it expelled many centrist members who could not see themselves voting for a hard Brexit. Encouraging this fractionation in society and in political parties is the reprehensible uses made of social media by domestic extremists and nefarious governments.

Social media, which is now prolific throughout the world and used to invade the psyche of competing nations and spread false information, creating false consciousness and thereby manipulating social and political thinking, has had great impact as the US example presented earlier in this chapter reveals. The invasion of one country into another's political and social affairs is not new, but what is new is the extensive use of social media in this effort. Social media is also replacing regular media, such as newspapers and TV, as the most predominant form of news consumed by many citizens, particularly the young. Social media is also a growing platform for activists to communicate with one another, speeding up the process of forming social actions and spontaneously organizing activities. A downside of this phenomenon is the bubbles created by social media activity. People become isolated in their own ideology and speak and listen only to those who think like them. This has led to further isolation in social discourse when, in fact, it should be opening the dialogue, not closing it down.

Political parties in the United States and other areas throughout the world are splintered and contain many partisan groups that are distancing themselves ideologically from one another. Each team is searching to become a social movement and control their party's direction. Much of this turmoil reflects the lack of consensus on a clear path to the future that will lift everyone up rather than leaving some behind. A clear path will become evident only when analysis of the present situation determines that a great leap forward in social and political thought is required to transform society into a new social and economic order much like what happened when society leaped from hunting and gathering to agriculture or from agriculture to industrialization. The question remains, can society come to a unified definition of the problem to be solved and then create rational policies to resolve it? Such a transition as the one being proposed here caused turmoil and distress in society in the past and is showing signs of doing so again.

There is also a major demographic shift in most western democracies that is having a dramatic effect on the ability of those countries to maintain their economic place in the world. Much of the turmoil in both cases presented earlier lies in the demographics of the country. The shifting demographics, where the aged are now the largest cohort in the history of modern western democracies, are having considerable impact on their societies. This phenomenon raises many issues that militate against what appears to be the needs and interests of those who are left behind in society. The number of people contributing to the growth of the economy is becoming fewer in comparison to those who are taking from it. Those on pensions, both private and public, are increasing, while those who contribute to it by working are becoming fewer by comparison. Corporations with defined benefit pension plans are seeking ways to slash their labour force or negotiate less-rich benefit packages and lower wages for new hires than they pay their older workers. The shift in demographics does not necessarily mean that countries who experience this phenomenon cannot continue to thrive; it simply means that these societies will need to take a leap in imagination and invent a new basis on which their society is constructed. It is time to develop and embrace a new social contract that takes account of the present circumstances and future problems of society in the technological and AI age. As the nature and makeup of society changes, so will its social architecture need to change. The stress we feel in society today is a result of a momentous change in social circumstance, but we struggle to make a corresponding change in the social architecture to cope with it.

An analysis of the Brexit spectacle and the election of Donald Trump to the presidency of the United States is very instructive in examining the dynamics of social upheaval and social change. The actions taken by both countries, albeit a misguided one, is an attempt to address the serious problems each country is facing. What path both have chosen is an attempt to retreat to earlier and better times. Whether or not they will be successful in their effort is yet to be determined, but one can speculate and suggest that attempting to go back to a period that no longer exists, or cannot be recreated, is not likely to advance their cause. The attempted retreat to the past must be seen to be only the first step, perhaps the first failed step, in creating a new social contract that will look and feel much different than anything that has gone before. So, let's not view these two stories as the concluding chapter of that book but just its introduction.

This chapter has examined recent experiments with social change from the political and sociological vantage point, and it can also be analyzed from the psychological perspective. This perspective can provide the overall analysis with more richness and focus to the individual experience than the other outlooks presented. When we think about the psychological perspective, we often gravitate immediately to psychoanalysis. This theoretical school asks us to look into the past for clues to our neurosis. It asks us to examine our early relationships when we were guided by an authoritarian figure (father and mother). The authoritarian figure protects us from the evils in society from which we cannot protect ourselves. Additionally, the same authoritarian figure is the sole provider of the necessities of life in our early years. This role can often be perverted at worst, or misinterpreted at best and can leave us with scars from those early years that continue to disrupt our adult life.

This relationship can play itself out in adulthood when social conditions become frustrating or appear to be overwhelming. It would appear to many who find themselves in a distressed condition that an authoritarian figure can protect us from identified threats from the outside world, such as declining security, both physical and economic (border walls and tariffs), and the fear of being swallowed up by an inscrutable culture (immigration). When these conditions intersect, the political terrain is open for nationalist and populist governments, perhaps leading to authoritarianism and even totalitarianism. The practice of psychoanalysis may assist society in explaining why such authoritarian figures as Donald Trump can become the leader of one of the most powerful countries on earth. This method of analysis can also be very useful in explaining why Britain chose Brexit instead of remaining in the European Union.

The recent election of Trump in the United States and the Brexit phenomenon provide us with an understanding of the power of fear and anxiety in a changing world. Although each of these examples is somewhat different in form, all are founded on the same concerns. They rely on the frailties of human anxiety and feelings of tenuousness. The last two chapters will provide some thoughts on how society can overcome these tendencies and move forward into the future.

Bibliography

Bremmer, I. 2018. *Us vs Them: The Failure of Globalism [Kobo version]*. Kobo.com.
Jeffko, W. 2018. *Contemporary Ethical Issues, 4th ed. [Kobo version]*. Kobo.com.
Kaplan, R. 2014. *Asia's Cauldron: The South China Sea and the End of a Stable Pacific [Kobo version]*. Kobo.com.
McClelland, J. S. 1996. *A History of Western Political Thought [Kobo version]*. Kobo.com.
Mueller, R. 2019. *The Mueller Report*. Washington, DC: The USA Department of Justice.
Wallerstein, I. 2004. *World-Systems Analysis: An Introduction [Kobo version]*. Kobo.com.

8
THE CONTEXT FOR SOCIAL CHANGE

The aim of this chapter is to examine the preconditions to, and context for, large-scale social change. When examining the great leaps in social change in the past, the most important question to ask is, can we identify the distinguishing social and environmental conditions that predict or facilitate social change, or is each transition so unique that it does not share features with previous events? If there are similarities, are these features displaying themselves today? For example, social change has been recently aided by the innovation of digital technology, especially personal computing, the internet, social media, and the worldwide web, providing instant communication across the globe. These most recent innovations in communication have, as in the case of the Arab Spring movement in the Middle East and North Africa, aided in the implementation of civil revolutions, some successful and long lasting and some not. Social and environmental context is a critical factor that can facilitate or force social change. I can't help but think that the state of our environment and the social injustice that exists in society today are threshold events demanding world-system change. Society needs to move from the present arrangement of the social system to how it *should* be constructed. This transition from 'is' to 'ought' will take a great leap of imagination and faith on the part of the political system and civil society. It will also take considerable effort on the part of all of us to determine what the 'ought' should look like and then create a path to get there.

As the reader may sense, moving from 'what is' to what 'ought to be' is filled with great danger. Infusing morals and values with verifiable science that predicts future events and consequences based on present behaviour may seem not only impossible, given the ontology of each domain, but also undesirable from some perspectives. While examining the 18th-century Scottish philosopher Hume's writings on the possibility of moving from 'is' to 'ought', Fukuyama declares, 'Hume believed that the "ought" and the "is" were bridged by concepts like "wanting, needing, desiring, pleasure, happiness, health" – by goals and ends that human beings set for themselves' (Fukuyama, 2002, p. 120). Not only do we set individual goals to guide our personal lives, but we also need to set collective goals to guide the development of the social world. There are many in society who will strongly object to what they would call 'social engineering' but consider the consequences of doing nothing. I have not attempted to describe in detail the various possible social and political scenarios that could develop if we don't abandon the present world-system that is heavily weighted by the predatory corporate capitalist system or what Harvey calls 'growthism', but it is a downward spiral in any case. For a rich description of the various scenarios that are our possible future if society does not act to rebalance the life-world with the system-world, I invite the reader to read Harvey's book entitled *Utopia in the Anthropocene: A change plan for a sustainability and equitable world* (Harvey, 2019). In his book, Harvey lays out in a brief chapter what is possible if we drift along without undertaking major change to the way we conduct our daily lives.

There has been great debate over the past about the morality of going from 'is' to 'ought' and for intervention in the social system, generally. While I don't dismiss these sentiments completely, I would argue that humans intervene in both the human and environmental systems regularly and have been doing so for some time. In fact, that may be what makes us modern humans, modern humans.

It is not possible to go from 'is' to 'ought' smoothly, and to suggest that it is a natural progression is a dangerous claim. 'Ought' implies some form of inevitable goal-oriented future. It injects a moral premise into what is an enterprise based on the scientific method grounded in present reality. Some would argue that this step is a 'bridge too far'. In the social sciences, however, this final step is required if we are to adapt society to ever-changing circumstances. The fact that we continue to create new technologies that affect the social and environmental systems regularly is a major part of what we humans do. If we continue to create and introduce new technologies into the physical and social environment, we will need to make adjustments or change completely the social environment in which we operate. Let's agree that we live in a purely created environment both social and physical and, therefore, have no alternative but to make large and small adjustments to harmonize with the planet and try to maintain some sense of homeostasis. It is the present lack of homeostasis that has brought us to where we are today and the intractable situation we face. The analysis of the present does not dictate what should be done in the future but provides clues to the possible positive and negative actions that are available. The future is not predicted but constructed based on what exists at present and the goals society wishes to achieve in the future. Some of these goals will address the inadequacies in the present system, so from that point of view, the 'ought' can be an outcome of what is in the present. Policy formation based on present science should clearly address the dangers and issues being encountered or be so constructed as to enhance the positive elements of the present. This procedure involves a process that identifies the values on which future society should be based.

Moving from what can be described as verifiable science and infusing it with socially constructed morals and values requires an enormous leap of faith on the part of society. The fact is we make that leap of faith every day, although most times based on individual ideological underpinnings and not on any systematic analysis of the present situation before us. It can be argued that most of us operate from a default mode, that is, we have constructed a predetermined way of reacting to the world. This way of addressing everyday life is methodically constructed during our early socialization process, and it allows us to be consistent in decision making without expending great energy in pondering every decision we make. The default mode is part of our unconscious mind. However, it may not be serving us well any longer given the complexity of the issues with which we are now confronted and the critical stage our society and environment seems to be experiencing at present. The default mode as a substitute for thinking and decision making based on verifiable evidence is part of our evolutionary history and appears to be the guardian of our short-term self-interest. It doesn't necessarily consider what may be in the long-term interest of the planet and the social system on which we depend for our continued existence, however.

Changes occur at many levels. Change at the societal level may be subtle in some ways and not in others. Societies are subject to changes through the periodic introduction of public policies or minor shifts in bureaucratic administrative direction that can slightly alter the trajectory of a local community or larger society but not take it down a completely different track. Many of the changes made at any level occur through technological advances that are introduced into the bureaucratic, production, or economic systems that make the sector in question more efficient but do not significantly affect the basic lifestyle of the citizenry or social architecture. Some interventions in the social system are serendipitous but have wide-ranging, long lasting impacts on society. That is not what is being suggested here. What may be needed in the present environment is methodical wide-ranging change that will set humanity on a new and more sustainable and socially just course of civilization.

Change at the ideological level can have profound effects on a single society and also extend to the world-system level. The French Revolution, for example, not only affected the French population but also inspired and transported the idea of democracy to many other parts of the globe. Some of these societies were engaged in what eventually became known as the Enlightenment. Was the French Revolution in any way a catalyst for the Enlightenment? Certainly, the philosophy of many French thinkers inspired the Enlightenment. It can be argued that these new ideas supported modern capitalism which became the basis on which the world economy functions, and nation states interact – in effect, the world-system under which we operate today. The notion of equality, the hallmark of the French Revolution, inspired other peoples and spread to many other parts of the world. Not only did it affect countless societies physically, as they adopted democracy as their political system, but also created and introduced new concepts into the world organization structure and the geopolitical system.

Historical scholarship suggests there are three possible catalytic events on which sociocultural evolution proceeds. The first is change *contingent* on other events in the environment. For example, changes to the sexual mores of society may be produced by technological invention and diffusion such as the development of the birth-control pill or an environmental event like the generation of a new disease. Certainly, one could argue that we are in the process of significant social change, contingent on climate change and the exaggerated inequality in wealth in our modern society, much as our sexual mores were dramatically affected by the advent of the birth-control pill and the human immunodeficiency virus. A more recent example can be found emanating from the COVID-19 pandemic. A large number of people were forced to work at home in many countries because of being quarantined to fight the disease. It is quite possible that after the crisis is over, working from home for many people will become the norm and commonplace. This is only one possible change among many that could result from the COVID-19 event.

In contrast to contingency theory, it is possible for there to be a *random accident* of stratospheric proportions that significantly changes the trajectory of society. For example, a nuclear accident could alter dramatically how nations produce and use energy or construct international relations in the future. Although modern society has not been subjected to a random accident, the proliferation of nuclear energy and its derivative 'the bomb' is a looming catastrophe given that many rogue states such as North Korea are on the brink of gaining that knowledge, or maybe possess it already.

Third is *conjuncture*. A conjuncture occurs in social relations when several independent developments or events converge at a certain point in time and interact with one another to produce a major change in the social or physical environment. For example, the French Revolution was a response to several factors, including the disenchantment of the peasantry for the oppressive aristocratic structure of society, the emergence from many years of external and civil wars, problems with financing the country, and the suffering and starvation of the French peasant population in contrast to aristocratic opulence. There may be a conjuncture of events evolving in the world today that could act as a catalyst for world-system change. In addition to the COVID-19 pandemic, the issues outlined in Part 1 of this book may constitute the independent elements that collectively provide such a conjuncture.

I have argued here and elsewhere that whether we like it or not, the capitalist world-system on which societies across the globe operate is entering a period of major change, and the question we must collectively address is what final form might we want it to take. This, of course, assumes that the human community can create a world-system rather than simply drift into one. It is likely that someone, or a group of people, will create a new world-system; it is just a matter of who that person or group will be, and the process used in its creation. Hopefully, it will be an inclusive process involving people at all social levels and not simply one that is driven solely by those with wealth and power who are dictating their version of social development at the moment.

The crisis of the COVID-19 pandemic has set the stage for the world to act to change the basic operating system. Will the COVID-19 virus act as a catalyst and motivate change to the present world-system, or will the world revert to the predatory corporate capitalist system unchanged after it has been subdued? We have witnessed the inadequacy of the present capitalist system to act positively in the face of the worldwide pandemic, so there is room for great change to be taken. In fact, the capitalist system itself could very well become a casualty of COVID-19. At the very least, predatory corporate capitalism has demonstrated its inadequacy to act positively in such a crisis. In fact, it can be argued that the predatory corporate capitalist system has taken precedence over the health of the people, and the oligarchy was willing to unnecessarily jeopardize lives to restart the economy. In addition, society has been forced to devote as much effort and resource to maintaining the capitalist system as it has deployed to fight the virus. There are forces in society that very much want to protect the status quo. In fact, during the crisis, President Trump more than hinted that a return to the full-blown economy takes precedence over the health of the nation by suggesting eliminating the quarantine precipitously and opening up the economy prematurely, perhaps putting the country and many frail citizens at risk again. He has suggested that the measures taken to combat the virus may be more harmful than the disease itself. This is another clear indication of just how out of balance the system-world is with the life-world.

Macro change is largely context driven. The historic method of spreading social change throughout the world in the past has been through innovation and diffusion. An historic example is the advent of agriculture and the storing of crops for future use. In more recent times, inventions such as the wheel or steam power are prime examples of sizable change in social relations. Recently, change has come from the social actions by citizen groups, such as the French Revolution from 1789 to 1799, the Cuban Revolution of 1953 to 1959, and the US civil rights movement in the early 1960s. A shared feature demonstrated in each of these movements was the strong leadership that inspired a large segment of the population.

Also unique in today's context is the declining liberal democratic political system that has abdicated its responsibility for creating and implementing social policy. Since the time society abandon the charity model of social welfare in favour of government social policy provision, most people have come to feel a part of democratic society. That may no longer be the case. Democracy is under attack. In democratic societies, the construction of social policy has not traditionally been left to the dictates of private enterprise or philanthropic institutions but led and preserved by the government. That tradition has been abandoned. Numerous long-held public institutions are being allocated to the capitalist market. For example, many institutions of incarceration have been turned over to private hands or are being staffed by private corporations. Taking someone's freedom away and handing their supervision over to the private sector raises many moral questions. Some would say it is anti-democratic. Other public institutions are becoming privatized as well. The charity model of social policy provision was previously abandoned by democratic governments, but a goodly portion of social policy was turned over to the religious community during the tenure of George W. Bush. Food banks operated by philanthropic institutions were to be a temporary measure and only operative until governments were able to mount policy to eradicate poverty. They have now become a permanent feature on the landscape. As we move toward a new world-system, the role of government in social policy provision will require rethinking and probably the reversal of present trends. It is becoming more evident every day that the ability of the private sector to produce a sustainable and satisfactory life for all citizens has been overestimated and the needed contribution from the public sector severely underestimated. The need for a modern social policy lies at the heart of world-system change.

The preferred option for making changes to the social architecture is through planning and policy reform. Reid provides a definition of social policy:

> (i)n its most basic form, social policy is about making choices and the processes involved in making them. These choices and the processes by which they are made rely on identified social purposes and the various ideological ideas that lay behind alternative options. Social policy is also about resolving conflict inherent in decision making by, in the first instance, including all segments of society in the decision-making process through some form of citizen participation. Social policy construction is fundamentally about engagement and inclusiveness in addition to the resolution of substantive social and individual problems in society.
>
> *(Reid, 2017, p. 13)*

Social policy construction is also about informal education through collective social learning. Social learning is mainly characterized as learning by doing. It is a social policy in itself, in addition to a means of reaching the end goal of producing a new world-system. It denotes intensive dialogue involving those who have a stake in the project. Social learning engages all the participants in the project equally, and all learn from one another rather than being educated by the elites or professionals solely, although professional input can be sought if needed but not placed in control of the process. It treats all participants in the process as equals, and decision making is bottom up, not top down. Social learning emulates the process associated with biological evolution. Waldrop tells the story of how Holland, one of the premier researchers in the field of systems theory, explains it. Social learning is based on Darwin's concept of evolution, which suggests:

> [A]n agent can improve its internal models without any paranormal guidance whatsoever. It simply has to try the models out, see how well their predictions work in the real world, and – if it survives the experience – adjust the models to do better next time. In biology, of course, the agents are individual organisms, the feedback is provided by natural selection, and the steady improvement of the models is called evolution. But in cognition, the process is essentially the same: the agents are individual minds, the feedback

comes from teachers and direct experience, and the improvement is called learning. . . . Either way says Holland, an adaptive agent has to be able to take advantage of what its world is trying to tell it.

(Waldrop, 1992, pp. 351, 252)

What the forgoing tells us is that system change through social learning relies heavily on constant feedback between the system and the environment in which it exists. It is heavily dependent on trial and error. Social learning is about learning from one's own lived experience and from the experience of others. Systems theory engages heavily in trial and error, and a major part of that trial and error is social learning that relies deeply on constant feedback among the parts of the system and the system in relation to its environment. These interactions make system change complex. Regardless of that complexity, people often organize themselves and engage in the critique of their societies and create organizations dedicated to educating themselves and their neighbours about the state of their life- and system-world.

Today's society requires a more sophisticated and mature platform for engaging social change than it did over the past. Historically, citizen movements were the most likely mechanism for producing large-scale social change, as in the case of the French Revolution. More recently, actions such as the civil rights movement that produced the Voting Rights Act of 1965 in the United States was a joint effort between a citizen movement and the government in power. The civil rights movement in the United States gave inspiration to other social justice projects throughout the world. What was initially required to start the process was for concerned individuals to identify the fundamental problem, demonstrate enough support for change, put heavy pressure on government to act, and then work with government to legislate the required policies and laws. This type of cooperative action between government and citizen groups may provide a modern replacement for bloody revolutions that occurred throughout history. This claim does not dismiss the recognition that the civil rights movement in the United States had its violent moments. Many civil rights activists were beaten or brutally murdered during that time. That said, hopefully we have learned from the past and will avoid bloody conflict during the next transition of society, but, given the rhetoric today, there is no guarantee.

The marginalization and isolation of many citizens in the capitalist system is increasing. It is occurring in a social system that promised, through hard work, to economically elevate everyone in society. The 'American Dream' offered to all in the United States has been a dismal failure for a large section of the population. The same disenchantment is occurring in many other parts of the world as well. This is certainly one of the factors contributing to the strength of Brexit. Those who suffer from this propaganda failure have become economically and politically marginalized in America and elsewhere in the world and may feel socially isolated in their own society. They no longer have a sense of belonging even though it has been promised to them repeatedly. Attaining some measure of social solidarity can lead to a positive sense of self. Unfortunately, social solidarity seems to be decreasing, not expanding, and many people are turning to aggressive actions to alleviate their fears, anxieties, and frustrations.

The new world of digital technology and robotics that requires a highly educated and skilled workforce is not accessible to many of those who flourished in the manufacturing economy, so employment is often difficult to regain for those who find themselves under-skilled and undereducated. This too has made the achievement of social solidarity less attainable, which, in turn, has led to the reduction in a positive sense of self and feelings of social belonging for many individuals in modern society. The reduction in levels of social solidarity has led to the rejection by many of society's institutions and the leadership of elites who have often overseen these foundational underpinnings of the social system. The social architecture no longer works for a growing number of people. The postmodern human is left with trying to find new avenues for achieving a place in society, and the road to that state appears to be very bumpy. The retreat to tribalism and rising levels of anger in the population is an attempt to regain a sense of self-worth.

To understand the preconditions for social change, we can start with the changing circumstances of the individual in society. The turmoil in western democracies today may be directly traced to the lack of a clear path for upward social mobility, a promise often made by democracy. While the capitalist economy may still be attempting to provide people with many of their basic needs, digital technology and robotics are having a profoundly negative effect on that attempt. Schumpeter cites Marx's idea of the 'formulation of the feelings of the unsuccessful many' (Schumpeter, 2016, p. 18) to describe the present phenomenon. The working class today demonstrate the frustrations that come from the unsuccessful many. In addition to their lack of skill to join the digital age, it is not clear that the new technologies will require the size of the workforce the industrial age demanded. The unmet psychological and social needs of the workforce are leading to the breakdown of society and the rise in social movements, such as nationalism and right-wing populism that make big promises but do little to address the fundamental issues contributing to the problem. The politicians promoting these ideologies may be providing emotional comfort to those left behind by prolonging the attainment of a positive sense of self-worth and social solidarity through their inaction on the real problems confronting society today. The restlessness in society will continue until better solutions to the problems produced by dislocation in the new economy can be found and implemented. The inability of nationalist and populist governments to sufficiently address the needs of the populace may lead to continued social anomie. Before instituting policies or, in Trump's case, tweets that supposedly substitute for clearly articulated policy, the actual problem that is causing the crisis needs to be identified and collectively agreed to by the larger society. Rather than talking about making America great again through attempting to recreate the industrial economy, western democracies must focus on developing alternatives to the workplace for providing the psychological and social needs that have been lost.

The COVID-19 crisis has shown us that there are many segments of society that have their own unique difficulties that need to be addressed by society.

A one-size-fits-all approach to governing or policy development does not work in a society that is multidimensional and pluralistic. For example, a society where work is the only avenue for gaining income is no longer tenable. Society will need to develop new methods, perhaps one like the guaranteed basic income, for providing a more equitable share of the resources to the population of the country and world. If we have learned anything from the recent pandemic, it is that governments have an essential role to play in safeguarding the welfare of the people they serve. Until the pandemic, governments in democracies had been vilified and made to appear irrelevant in favour of extolling the questionable virtues of the free market. The COVID-19 virus made clear once again that governments have a substantial role to play in engendering social solidarity and providing basic services to the population.

The social context at any given time will be a contributing factor for the type of change that is sought. If the basic needs of society are threatened or weakened, change is likely to be motivated by fear or anger. Society will probably concentrate on regaining what appears to have been lost, and its focus can be expected to be deficit driven. On the other hand, if the social atmosphere is positive and social solidarity is present at some level, society will pursue social growth, analogous to the individual need for self-actualization as described by Maslow (1943). Each of these conditions can be seen in the slogans of two recent US presidents. Trump's 'Make America Great Again' slogan is clearly a deficit-driven jingle. Deficit-driven motivations are built on loss of social solidarity. Obama's 'Yes We Can' slogan is a social growth motto that is socially analogous to reaching the state of self-actualization by individuals. Increasing social solidarity is the gratification of a need for personal growth and social development as envisioned by Maslow (1943).

Striving for social growth requires transition to a higher state, built on a 'Yes We Can' social climate. The new world-system will require a society that is seeking (society-wide) social and political actualization, the motivation that is personal and social growth oriented, not deficit driven. To achieve a new world-system, politicians and politics will need to retreat from 'Trumpist' dogma, which promotes fear and anger in the population, to one that is positive in its outlook and views the challenges in the world to be uplifting and potentially developmental and not a zero-sum game. The alteration of the social and political atmosphere in which we live today is a critical first step in laying the groundwork for world-system transition. Much of this atmosphere is created at the highest levels in government. Certainly, there is next to no chance of moving to an appropriate new world-system within the social atmosphere that exists today. Changing the political and social discourse from being divisive and negative to positive and future oriented will be the first major challenge and step on this journey. Social climate and context are very important when attempting world-system transition. Change relies on opportunity. Maybe the COVID-19 pandemic provides such an opportunity.

Changing the social climate will require a concerted effort at all levels of society. It cannot be left simply to chance. And it needs to be recognized that there are strong forces in society who find the present social climate conducive to their interests and they will take strong measures to maintain the present social atmosphere. Believe it or not, some people thrive, both psychologically and economically, on a negative social climate, so there will be resistance to change. Certainly, the present social climate is palpably divisive and one that thrives on blaming others for their own inadequacies, which will not be conducive to massive progressive change.

Societies make small and large changes continually to their social systems and in a variety of ways. Regardless of whether those changes are big or small, they have the potential to dramatically affect lives and so are regularly resisted. Human culture is constructed so that behaviours that have been useful in the past are passed on from one generation to another, and those practices are very difficult to abandon in favour of new ones that may not have demonstrated their worth yet, even though their adoption may seem rational. In fact, human cultures have a strong built-in inclination to defend themselves rather than engaging in radical change. However, given the environmental and social chaos confronting humanity today, radical change is very much needed. More will be said about working at the community level to bring about that change in a later section of this book, but suffice it to say here that the quest for a sustainable environment, the need for social justice, and an equal voice for all citizens in the political domain, are the first steps in satisfying the basic needs of society and in laying the framework for focusing on increasing individual self-worth and social solidarity.

Dramatic and significant change occurs in the form of a social movement and not necessarily as an initial undertaking by government. Mature capitalist economies like any system or function can become stagnant – even corrupt – over time. This assertion is particularly true at the political level. At the very least, politicians often have difficulty separating out what is in the interest of the country they serve and their own interests. Supposedly, the democratic system is designed to rectify this problem at the voting booth where informed citizens can replace politicians who are seen to be operating contrary to the best interests of their constituents or country. While theoretically this is the tradition of the democratic system, there are many intrusions into this process that militates against this practice. The injection of unlimited amounts of money into the political system has prejudiced communications (some might call it propaganda) and the campaigns of politicians and has severely distorted the way the system is supposed to work. It is becoming increasingly difficult to undertake and complete social change of the magnitude discussed here because of the isolation in the political system that big money has created. As a result, it will take bottom-up grassroots organizations, not much different than the civil rights movement of the 1960s, to achieve the magnitude of change that is suggested here. More will be said about the construction and role of grassroots organizations in world-systems change in a later chapter of this book, but let there be no mistake, it must form the basis for achieving the type of change discussed in this volume.

Changing the social architecture does not transpire in a vacuum. All world-systems are set in a context. While there is no known formula for altering the status quo, the social circumstance at the time of contemplated change will be an important factor in reaching success in the end. There must be a readiness for change. The turmoil we witness in society today suggests there is a threshold for change at this time. However, the divisiveness that exists in society will not be conducive to the type of wide-ranging change that is required to address and rectify the dysfunction that exists in society. People interested in social change need to focus on the atmosphere in which that change is to occur. There needs to be laser focus on unifying all society to the cause that is being contemplated. An atmosphere of dialogue and trust in the system needs to be created to aid system change. As an aid in moving toward change at the system-world level, theories that describe and detail systems and system change are useful if not critical. A review and analysis of the most important change theories that can help guide world-system change is set out in Chapter 9.

Bibliography

Fukuyama, F. 2002. *Our Posthuman Future [Kobo version]*. Kobo.com.
Harvey, M. 2019. *Utopia in the Anthropocene: A Change Plan for a Sustainable and Equitable World [Kobo version]*. Kobo.com.
Maslow, A. 1943, July. A Theory of Human Motivation. *Psychological Review* 370–396.
Reid, D. G. 2017. *Social Policy and Planning for the 21st Century: In Search of the Next Great Social Transformation*. London, UK: Routledge.
Schumpeter, J. A. 2016. *Capitalism, Socialism, and Democracy 2nd. ed. [Kobo version]*. Kobo.com.
Waldrop, M. M. 1992. *Complexity: The Emerging Science at the Edge of Order and Chaos [Kobo version]*. Kobo.com.

9

THEORIES OF CHANGE

This chapter presents and explores social theories that help explain the idea of social change at the world-system level. Its intent is to sketch out theories that emanate from a variety of academic disciplines to accomplish two goals. First, to enhance the reader's understanding of the phenomena under investigation and, second, to provide theories that can be instrumental in guiding the change process. Most of the theories presented throughout the remainder of this chapter will accomplish both tasks. The theories discussed can be integrated into a framework for undertaking a process leading to world-system change.

A theoretical foundation is critical to the social change project. Without theory, humans simply march forward relying on intuition but not on reasoned rational goal setting based on the assessed needs of society. Theory provides the blueprint for creating a possible future. It can predict how the social system and environment will react to change. Any long-lasting change must be grounded in our best ideas and models about society and the environment, and we must not march into the future blindly and without the guideposts theory provides.

Uncovering or creating transition theories that explain the processes leading to world-system change has met with limited success. What may be difficult in achieving such a proposition is the level of focus by present-day scholars. Most social science practitioners focus on micro and mid-level (mezzo) society, but the capitalist structure is truly the first all-encompassing world-system that has reached all corners of the globe. Micro or mezzo level theories may not lend themselves to analysis at the macro scale without some adjustment. World-system change, however, requires focus at the highest level. Micro and mezzo scale theories that show some potential for this task may need to be recalibrated given the difference in magnitude of the goal of this analysis.

The Enlightenment that began in the 18th century initiated the modern world's march to the scientific, technological, and industrial era. This was the beginning of the overall social complexity that we experience today. The Enlightenment paved the way for capitalism. There have been many attempts down through the ages to create theories that explain the motivating factors for social change and how new eras are ushered in and old, worn-out cultural practices, left behind. In addition to the thinkers who ushered in the Enlightenment (Hobbes, 2012; Locke, 1689; Rousseau, 1988), Adam Smith and Karl Marx (Marx, 1867; Smith, 1771) explained the capitalist economic model that was emerging at the time and then modern social observers such as Galbraith and Giddens (Galbraith, 1984; Giddens, 1987), among others, articulated macro theories on industrialism and the administrative society, which explained the organizational instruments that were necessary for supporting the development of industry, the backbone of the capitalist system. Industrialism was the mechanism that replaced the cottage industries and guilds, dramatically expanding the amount and type of goods that could be produced in less time than it took using earlier methods. As time went on and the system matured, people benefitted from this advance as they became wealthier, obtained more nutritious food, and material goods. The inventions of new technologies and the division of work in the manufacturing process allowed the Industrial Revolution to take hold. New methods of energy production, initially the harnessing of steam and then the combustion engine, increased the effectiveness of the industrial model. The innovation of electricity changed individual life dramatically. Growth in wealth generated by these new methods of production increased the living standard for most people in society. That said, there have been people left behind and marginalized when society transitioned from one social system to another. This unfortunate fact is proving true for modern society again as the world engages in new biotech, digital technologies, AI, and robotics.

Because of the rapid expansion of production in the industrial era, huge amounts of new capital were needed to facilitate development and growth. The pooling of capital by the banding together of wealthy merchants with the landowning classes created the corporation and initiated an ownership class – the capitalist class. Early on, investors were sought to finance these companies to spread the risk rather than having it solely underwritten by a single individual. Eventually, many of these companies became so large that operation by a single individual became almost impossible, although not unheard of, and the administrative class arrived on the scene.

The rise of the management class has been well documented by John Kenneth Galbraith in his 1967 classic work titled *The New Industrial State*. Today, there is often a separation between the capitalists who invest in the public corporation and those who run it. The managerial and administrative class has become a considerable force in the capitalist era, and at times their goals and those of the shareholders conflict (Galbraith, 1967). This is the social space the capitalist world occupies today.

The reminder of this chapter identifies and analyzes theories that help explain the motivations for, and processes of, social change. Any attempt to understand and facilitate system change will depend on a solid theoretical foundation. Theory provides us with a construct of the phenomenon under study. Investigators can build on original conceptions or refute them completely as the investigation proceeds. This chapter will not only explain change but also provide potential frameworks for stimulating change at the macro scale. Change and adaptation are pursued by agents acting, hopefully, in a logical and rational manner even though they may be initially driven by emotional (value) intelligence. One basic concept useful to this project is heuristics.

As an overarching analytic methodology, heuristics offers an approach to problem solving that takes a practical and expedient approach to understanding present reality and in reaching a predetermined goal, such as world-system change. Heuristics aids in problem solving, learning, and discovery and can be employed to understand or enact change at the world-system level. It facilitates exploratory problem solving and embraces the concept of trial and error as well as experimentation in problem solving. Many of the approaches to heuristic understanding is founded on the principle of experiential research and learning. Change as large as world-system change will require extensive trial and error as we discover what enhances life and social cohesion within a new social framework and what does not. Systems analysis and action research are heuristic methods useful in pursuit of the goal set out in this book.

System analysis, chaos, and complexity theory are appropriate for synthesizing large amounts of research data to discover patterns in those data and conceptualize it as a system. Action research is applied on an ongoing and repeating basis when alternating between data collection and analysis, and implementing actions based on those data. Reductionist research methods are suitable for understanding components of a single variable by determining the constituent parts and actions of that variable. All three approaches to research and understanding will be useful to the focus of concern here. Reductionist methods are best used for understanding the issues outlined in Part 1, and heuristics are suitable for the investigation undertaken in Part 2 of this book.

Emotional intelligence and rational choice have aided survival of humankind at various times throughout history. Humans are programmed to react emotionally to threatening situations, even false positives, as safety and defence mechanisms. Responsive programming motivates society to act when threatened based on a clear set of human values handed down from one generation to the next. That said, events that trigger emotional response today are much more abstract than during earlier periods of human existence. As a result, solutions based solely on emotional reaction may not result in an adequate or useful response. Today, humans must combine emotional intelligence with rational choice analysis to arrive at solutions that will provide success in resolving the present crises in society.

Just exactly what is one to look for in guiding the transformation of society? What is required here is not simply reworking the present system at the margins but a total transformation of society – a transformation as large or larger than the great leap from hunting and gathering to agriculture, or from agriculture to industrialization and the scientific revolution. Social science since the days of Marx has been reluctant for whatever reason to focus its effort at this level of enquiry. Wallerstein suggests, '(w)orld-systems analysts see themselves therefore as engaging in a fundamental protest against the ways in which we have thought that we know the world' (Wallerstein, 2004, p. 9). This does not suggest that social science has not had a vital role to play in attempting to ferret out some understanding of how systems change; it is often the unit of analysis and hence its limited explanatory value at the macro level that is in question. There will be the need to determine the structures and, most importantly, the processes required to drive any eventual social alteration. A large part of sociocultural development is accomplished by motivating large groups of people through emotional and rational argument. Rational argument would be preferable to pure expressive effort, but history would suggest that humans are emotional creatures, and many individuals see the world through emotive eyes as evidenced by the power of religion in conducting human affairs down through the ages. That said, planning based on fact and reason must inspire this project given its intricacies and complexity, to say nothing of its importance.

When speaking of reason, Jeffko suggests there are ' . . . three basic modes: *instrumental, aesthetic*, and *moral*' (Jeffko, 2018, p. 36). I understand instrumental rationality to mean motivations that pursue an individual utilitarian result. The result may be damaging to the physical or social environment, but it serves the ends of the actor, and that is the motivation for acting. Aesthetic motivations, on the other hand, are those inspirations that are stimulated by personally held values or principles of the group or tribe. Neither instrumental nor aesthetic values necessarily take account of the moral consequences or the outcomes produced by the action on the larger physical or social context. Neither are directed by nor incorporate outside forces such as science or expert opinion as the foundation for acting. Moral or ethical rationality, however, is cognizant of fact-based physical and social science. Ethical rationality is concerned not only with achieving outcomes that are personally satisfying, either from the instrumental or aesthetic point of view, but primarily concerned with what may be in the best interest of the planet and the social system.

Further to the various categories of rationality enunciated above, reason is determined by social action as much as it is by individual thought. People in dialogue come to determine what is rational as much as any individual acting or thinking on their own. The latest thinking on the condition of rationality incorporates the notion of –

> 'bounded rationality', which means that we have biases and cognitive limitations that prevent us from realizing full rationality. Our dependence on social interactions, however, isn't simply a bias or cognitive limitation. Social learning is an important method of enhancing individual decision-making. Similarly, social influence is central to constructing the social norms enabling cooperative behavior. Our ability to survive and prosper is due to social learning and social influence at least as much as to individual rationality.
> *(Pentland, 2014, p. 532)*

Instrumental and aesthetic rationality are the dominating rationalities of present-day society. Politics throughout the democratic world has become very polarized and is now being characterized by some pundits as tribal. Ethical or moral rationality has become subservient to instrumental and aesthetic rationality, perhaps even absent in today's politics. The various groups or tribes that exist in society today are now locked into their own aesthetics, and their ideologies will be extremely hard to change without a well-constructed and defined process for dialogue. Sectors of society are now chasing what is instrumental in achieving their aims and those goals do not appear to take account of the larger good. This way of acting in society presents a major stumbling block in achieving a renewed world-system and a resolution of the problems identified in Part 1 of this volume. In fact, it may be that there is no consensus on exactly what problems need addressing. Consensus on this issue may be one of the first hurdles to overcome during the process of system change. Emotion or aesthetic rationality is a unique human characteristic and has served humanity well in the past, but its dominance in decision making may now need to be replaced with greater reliance placed on rational ethical analysis and management. All actions proposed through instrumental or aesthetic rationality must be subsequently viewed through an ethical lens.

The traditional orthodox scientific method is dedicated to determining the structure and function of the phenomenon under study. It accomplishes this task by reducing a phenomenon to its constituent parts to discover their fundamental elements and how they function. While reductionism is important to the discussion here, particularly for understanding the issues raised in Part 1 of this book, it is not suitable when those matters are viewed as a systems problem. Given the purpose of this book, the mission requires a strategy that is synthesizing and expansionary rather than reductionist. We are interested not only in the functioning of each part in the system but also how those parts relate and interact with one another in the larger whole. How change to one element in the system affects all the others is critical for understanding world-system change.

General Systems Theory (GST) is concerned with the relationship between the parts of a system and not necessarily with the functioning of each individual part in isolation to the rest. Ludwig von Bertalanffy (1956) and Kenneth Boulding (1956) were prominent in developing the concept of GST and advocated for its role in science during the 20th century. It is worth quoting Skyttner on the differences between reductionist analysis and GST. Simply stated reductionist

> analysis gives description and knowledge; systems thinking gives explanation and understanding. . . . GST provides a way to abstract from reality; simplifying it while at the same time capturing its multidimensionality. As an epistemology it structures not only our thinking about reality but also our thinking on thinking itself. Systems thinking is a response to the failure of mechanistic thinking. . . . Systems thinking expands the focus of the observer, whereas analytical thinking reduces it. Whereas analytical thinking concentrates on static and structural properties, systems thinking concentrates on the function and behaviour of whole systems. General Systems Theory was founded on the assumption that all kinds of systems (concrete, conceptual, abstract, natural or man (sic) made) had characteristics in common regardless of their

Theories of change **147**

internal nature. General Systems Theory is, however, not another discipline – it is a theory cutting across most other disciplines linking closely e.g. generalize concept of organization, to that of information and communication.

(Skyttner, 2001, pp. 31, 36, 37)

Systems theory provides an obvious method for approaching a subject as large and multidimensional as change to the world-system. Each of the parts in the system, as outlined in Part 1 and in Figure 9.1, will respond to how each of the other parts in the system are constructed and manipulated.

The problems (attractors, in chaos/complexity theory jargon) that constitute Part 1 of this book were subjectively but carefully chosen. In a perfect data-rich world, the problems selected to constitute the system might have been selected using a statistical method such as cluster analysis. Given the lack of such a complete data source at the world-system level, the items carefully chosen here were selected through a rigorous secondary analysis of the qualitative and quantitative literature provided by many experts esteemed in their subject fields. Much of that literature is summarized and presented throughout this book, particularly in Part 1.

Most systems studies are retrospective in nature. They ask the question, has the system changed over time, and, if so, what changed it. The subject of this book is quite different. It is not concerned with determining what change has taken place retrospectively but how we can make desired changes to the present world-system or, more likely, create a new one. This orientation does not lend itself to the statistical manipulation of data but endeavours to construct a range of scenarios that have the potential to change the system positively in order to reduce chaos and find order. The goal here is to design the future, not predict it. Data gathering and analysis, using such methods as attitude surveys, content analysis of secondary data, and other such data-gathering methods, can find use in the inclusive process of

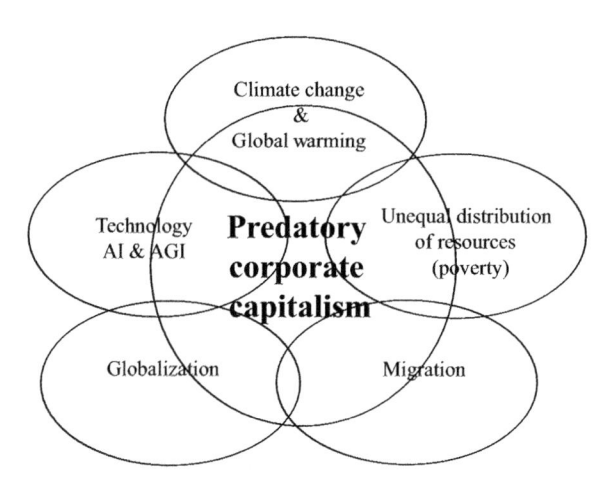

FIGURE 9.1 The world-system and component parts requiring alteration

system change where agreement regarding approaches to, and direction of change in the system is the goal. World-system change is aspirational, not predictable, and therefore requires a much different approach than studies that attempt to define the system and its historic or present behaviour.

Perhaps the most fundamental question to be answered when thinking about systems and the theories that explain them is, what exactly is a system when we are concerned with the social? Byrne takes this question head on.

> I have been troubled by the issue of whether the social constitutes a system or whether the social is the space within which that system is located. The only answer I can come up with is that it is both. . . . at one level of analysis social systems are located as entities with states describable by co-ordinates in n dimensional spaces.
>
> *(Byrne, 1998, p. 41)*

When change occurs to one or more variables in the system, we cannot predict exactly what will happen, but we know that change will be contained within the social space and time in which it exists. As Byrne tells us,

> [S]mall changes through time produce indeterminate results: anything could happen. The interesting thing about complex solutions is that we can't predict what will happen, but we know that what will happen will be drawn from a set of alternatives greater than one but less than too many to cope with – the realm of determined chaos.
>
> *(Byrne, 1998, p. 41)*

Certainly, the alteration of one or more variables will affect the others in the system and change the configuration of the system itself. To be clear, we are interested in the composition of the components of the system, but also in understanding how the system itself acts and reacts over time. What happens to the system as it responds or not to the various changes in the parts that comprise it or by alterations to the environment in which the system exists? Systems analysts also concern themselves with the hierarchical relationships among the data as well as the multiplicity of actions. Can we gain an understanding of the importance of each variable to the overall system? As alterations to the various elements in the system are introduced, we need to be sure not to negatively affect the other elements in the system or the system as a whole.

The concept of system can be useful in a number of situations from big to small. At the very base is the idea that the whole is greater than the sum of its parts. Skyttner emphasizes:

> (c)omplex systems are characterized by a large number of elements; many interactions between the elements; attributes of the elements are not predetermined; interaction between elements is loosely organized; they are

probabilistic in their behaviour; the system evolves over time; and subsystems are purposeful and generate their own goals; the system is subjected to behavioural influences; the system is largely open to the environment.
(Skyttner, 2001, p. 98)

Skyttner's conceptual description of a system encourages us to view the individual parts that operate independently in a closed system as dynamic entities that interact with one another, changing their shape and function over time. The relationships and actions among the parts create a dynamism that is chaotic and continually changing. In complexity theory,

> [T]he basic components and the basic laws are quite simple; the complexity arises because you have a great many of these simple components interacting simultaneously. The complexity is actually in the organization – the myriad possible ways that the components of the system can interact.
> *(Waldrop, 1992, p. 117)*

Examining the interaction of the parts as a dynamic system is more complex than analyzing them to discover their individual workings in a reductionist manner.

Systems thinking also places emphasis on the planning and management of change. It does not countenance the idea that humans cannot create reality but simply discover it.

> (d)esign or redesign becomes the key concept of the systems perspective when it is about to change the world for the better by building new or improved systems. Model building and optimization replaces guesswork. It is concerned with how things ought to be, with combining resources to obtain goals. Systems design (or systems synthesis) is a formal procedure where human resources, artefacts, techniques, information and work procedures are integrated into a system in order to facilitate performance.
> *(Skyttner, 2001, p. 40)*

Augmenting systems thinking in the quest for understanding are the notions of chaos and complexity. Complexity theory concerns itself with what is occurring on the edge of chaos. 'Complexity was defined as the domain between linearity determined order and indeterminate chaos' (Byrne, 1998, p. 8). Its focus is on the complex process of system adaptation. It is multidisciplinary. It understands that the real world does not operate according to exclusive disciplines, as established in the academy, but in a messy and integrated fashion. Its focus is on the spaces between the disciplines (Waldrop, 1992). Those employing chaos and complexity theory do not operate from a single discipline but interact and mingle. Individuals working in this new field,

> [B]elieve that they are forging the first rigorous alternative to the kind of linear, reductionist thinking that has dominated science since the time of

Newton – and that has now gone about as far as it can go in addressing the problems of our modern world.

(Waldrop, 1992, p. 12)

Complexity theory is essential to understanding and contemplating world-system change. It centres on understanding complex forms and in clarifying the nuances of those discovered patterns. It basically focuses on the edge of chaos, and to quote Byrne, ' . . . chaos is the precursor of order, not its antithesis' (Byrne, 1998, p. 13). This is the exact space the present world-system occupies, and complexity theory focuses on the emerging order emanating from that chaos. It ' . . . deals exactly with non-linear relations, with changes which cannot be fitted into a simple linear law taking the form of statement of single cause and consequent effect' (Byrne, 1998, p. 13). It is important to note that it is not the chaos which is important to us but the transformation of it into a new paradigm, leading to order. And let there be no doubt, not only will that transformation be complex, but will also result in a new and vastly changed system.

This book is about understanding the present chaos and its complexity in society and moving to the next stage of social order. Complexity theory understands that large systems are dynamic and ever-changing and not static arrangements. Additionally, 'if we use this approach, then we can resolve the major issues for sociological theory of the relationship between macro and micro, and structure and agency' (Byrne, 1998, p. 18) as discussed elsewhere in this volume. In addition to the other theories presented in this chapter, chaos and complexity theory provide a framework not only for explaining the present condition of society today but also for creating a rational process for changing something as complex as the world-system from its present chaotic state to a more functional structure, taking humanity to the next level of sustainability and civilization.

What chaos and complexity theory gives us is the understanding that predictability in complex systems such as the world-system is not accessible and not the goal or object of study. Not unlike physical science, social science at this stage of its development may need to accept the notion of probability as its basis for theory construction and acceptance. Probability which includes an acceptable margin of error in its condition for acceptance can be very helpful in explaining the validity and reliability of the system under study. For example, projections made about climate change by many groups examining global warming employ a form of probability using a scale to estimate certainty, something akin to 'not likely' to 'highly likely'. Quantum physics routinely uses probability in its explanations, so why not social science? On the face of it, this approach seems like a reasonable tactic. Of course, the question that remains is mainly that of defining what constitutes probability. This definition may have to be driven by the context and subject under investigation. What may be most important in this regard is that the definition of probability be laid out in a clear and transparent way in advance of the research so that study evaluators can come to a determination of whether the theories being expressed seem reasonable and valid or not.

Chaos theory considers the idea that some of the parts making up the present system were random chance events created by their context and have come to dominate because of the awareness of the notion of 'increasing returns' (see Waldrop, 1992). Phenomena that dominate everyday life may not be the most effective way to accomplish life's events but are dominant because they have become 'locked in' due to a series of minor decisions in the past that became significant and non-changeable over time. They become non-changeable because other systems or parts of systems were built around them. Once locked in, these elements take on a life of their own and proliferate until they become anti-utilitarian. This is how society's distorted form of capitalism arrived at its preeminent position, not necessarily because it's the most effective system but because of decisions taken at an earlier point in time. This process has become known as the process of 'increasing returns' first considered by Stanford University economics professor Brian Arthur (see Waldrop, 1992). The idea of increasing returns is the antithesis to the idea of equilibrium, which has been long-standing orthodoxy in economics. Chaos and complexity theory accept the notion that systems are self-organizing and change through small alterations made to the constituent parts and their relationships that get locked in and become significant over time through both positive and negative feedback. These changes produce new configurations and patterns in the system. Change to the system can occur through small changes to the fundamental patterns of the system by altering the locked-in nature of its elements. Even the smallest changes to the individual elements in the system changes the overall system significantly over time.

The problems identified in Part 1 of this book act as attractors, the term formally used in chaos/complexity theory, that is, super-variables into which the remainder of the other variables in the system connect. Waldrop (1992) provides an analogy to describe attractors. He likens them to a pond surrounded by hills. When it rains, all the water runs down the hills in gullies into the pond making the pond the attractor, and the numerous rivulets streaming down the hills, variables feeding into the pond. As this action occurs, the composition of the pond changes, as does its shape. In fact, the pond is ever changing even when it may appear to be dormant.

What makes systems complex is the relations among the parts. A relationship of one part to another creates a third variable that must be considered, described, and explained in the equation. The connection of poverty to climate, for example, may produce a subsystem of its own made up of a myriad number of parts (policies), all of which have the potential to change the overall system. As you can see, the potential for there to be numerous subsystems associated with each variable, and additional subsystems within the secondary variables and so on, makes the outcome of small changes to the system extremely complex and indeterminate, particularly when a change to one part may cause additional changes in the others. As the analysis grows deeper, the system becomes increasingly complex. Variables are multiplicative, not additive as in the case of linear research, and each interacts with all the other parts in the system. Think of what happens when one of those parts

152 Processes of change

is changed in some way. That change will not only influence the other variables in the subsystem, and then the larger original system, but also change the shape of the entire organization itself. Once one variable is altered, the entire system changes its configuration. So, we may well have an altered configuration, the shape of which no one can exactly predict in advance. That is why systems analysis does not attempt to predict the future of the system but learns how the system reacts and adjusts to changes made to its many variables. It is often the case that systems theory deals in probability rather than predictability unlike reductionist science. Systems thinking focuses on pattern creation and alteration rather than cause and effect. It is this interaction among the parts in the system that necessitates addressing all the domains outlined in Part 1 as a whole and not as independently acting agents. The application of chaos and complexity theory to world-system change will cause us to contemplate how the alteration of each variable in the system will affect the other variables and the overall pattern of the system. Largely it is the agency that often makes the difference in which direction the system takes. This fact alone demonstrates how important it is to address all the problems outlined in Part 1 of this book collectively and not individually as some might want to do at first blush.

Following along from establishing the overarching role of GST and chaos and complexity theory in examining and promoting change to the present world-system, it is imperative to understand how large fundamental systems react to stress and distress. The diagram below outlines the major theories that explain the process for both the breakdown in the biological (Selye's GAS) and social world (Kuhn's paradigm breakdown).

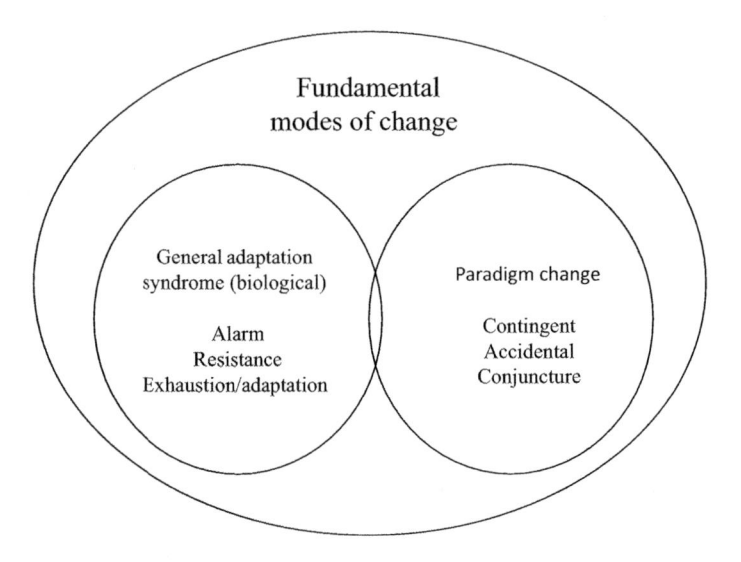

FIGURE 9.2 Theories of biological and social adaptation

Source: Hans Selye.

Fundamental to the story of social transformation is the life-and-death struggle that occasionally confronts specific biological species or social systems. The life-and-death struggle that we are familiar with not only applies to people and other life forms but also to social systems. If appropriate change in social patterns and practices evolve and respond successfully to changes in the environment, a culture continues, but if it doesn't, then life, in this case a social or cultural system, peters out or ends abruptly. Perhaps an appropriate but often missed place to start when considering social transformation is not from the social science perspective as much as from the biological and natural history literature. It is often useful to use a metaphor from an unrelated field to provide a conceptual picture for a complex process in the social sciences. Using Selye's General Adaptation Syndrome (GAS) provides a basic model from the natural sciences to examine social system adaptation and change. Additionally, the issues of climate change and global warming have everything to do with the physical environment and the biological creatures that inhabit it, including humanity, so Selye's theory provides a useful lens for exploring the world's reaction to environmental stress. Some social scientists may have difficulty with adapting a biological theory as a partial explanation for a social condition. Although I understand those misgivings, I believe that cultures and societies undergo a comparable process to the biological course that Selye's GAS suggests, at least from an analogous point of view. It is in that spirit that Hans Selye's GAS (Selye, 1956) is considered important to this analysis.

Selye's model suggests that an organism, when confronted by a stressor, advances through the stages of *alarm, resistance*, and eventually *exhaustion* and dies out if it is unable to resist the stressor or adapt to the stressing agent. If it does deal successfully with the stressor, it reaches a stage of adaptation and returns to homeostasis. The elegance of Selye's GAS theory is its simplicity. Like biological evolution, cultures and societies undergo many stresses emanating from the environment and must adapt or perish, and Selye's model reminds us of that eventuality. Social and cultural systems are dynamic, not static, and must overcome social stress through change. The world-system is not immune to that fundamental fact. Unlike the biological system, the social structure may not die out but enters a stage of chaos and dysfunction.

In the social milieu, adaptation or sociocultural evolution is the logical, and successful, outcome of resistance to a stressor. Exhaustion should only occur if the organism, or in this case the social system, cannot successfully overcome the stressor by transitioning to a new and more appropriate social structure. In present society's case, the stressors are global warming, increasing separation between the rich and poor, and increase of relative poverty throughout society, to name but a few (see Figure 9.1). The GAS may provide a useful metaphor for the process humanity seems to be addressing, or ignoring, at the moment. Unlike the biological system, the extinction of the social system will likely occur gradually and over time, not necessarily abruptly. What Selye's model provides this discussion is the realization that systems are not static entities but very dynamic and subject to extinction if the alarm and resistance phases are not addressed adequately.

154 Processes of change

Thomas Kuhn's work contributes to the understanding of social change even though his work is specifically focused on scientific paradigm change. Kuhn (2012), in his book *The Structure of Scientific Revolutions*, focuses on paradigm change in science. Kuhn's theory provides a conceptual explanation of the process of social change.

Like the processes of change envisioned by Selye, Kuhn outlines the change process that a physical science navigates as it uncovers new knowledge that puts into question the current orthodox ways of viewing that science. Kuhn explains change as an alteration or abandonment of the prevailing orthodoxy of the science in question, to a completely new way of thinking and viewing the subject. Science is essentially puzzle solving. All science at its core is constructed on a basic guiding paradigm that after time becomes unable to confirm orthodox theories in light of newly discovered data or new research methods. A new improved paradigm is required to explain the phenomenon. The new paradigm theorizes and examines the phenomenon in an entirely different manner, resulting in new insights. A paradigm shift provides a new way of looking at the phenomenon. As Hacking in the introduction to Kuhn's book explains,

> (a)ll is well until the methods legitimated by the paradigm cannot cope with a cluster of anomalies; crisis results and persists until a new achievement redirects research and serves as a new paradigm. That is a paradigm shift. . . . a paradigm is not only an achievement but also a particular way of modeling future practice upon it.
>
> *(Kuhn, 2012, p. 15)*

To put it in the context of the subject being discussed here, the present paradigm that depends so heavily on capitalism and market forces to explain much of humanity's social relations is no longer adequate to address the present condition of society. The problems today are not a matter of continuing to grow the economy but in social justice, equality, and environmental sustainability. To quote John Kenneth Galbraith,

> To furnish a barren room is one thing. To continue to crowd in furniture until the foundation buckles is quite another. To have failed to solve the problem of producing goods would have been to continue man (sic) in his (sic) oldest and most grievous misfortune. But to fail to see that we have solved it, and to fail to proceed thence to the next task would be fully as tragic
>
> *(Galbraith, 1984, p. 259)*

Kuhn was speaking and writing about physics because it was his discipline. He was a physical scientist. However, his ideas are just as applicable to the social sciences as they are to the physical sciences because they are about human processes, not scientific subject matter. He understood that theory was constructed over time by people adhering and working in agreement on an orthodox process for theorizing, collecting, and analyzing data in a particular discipline. This is what I understand him to mean by the idea of paradigm. Kuhn wants us to understand that what he calls 'normal science' operates until an impasse, or 'anomalies' are encountered at which time a new paradigm is offered to replace the old one. Take, for instance, the Copernican Revolution that completely changed how we view the universe. Consequently, schools of thought sometimes antagonistic to one another develop inside a discipline. We witness the same breakdown occurring in the general society today. Because the formation of a paradigm is anything but benign, Kuhn referred to this process as a scientific revolution. As Kuhn suggests, theories develop by the accumulation of knowledge and then are discarded or changed with the creation of new knowledge, or the anomalies that cannot be explained by the orthodox theory. Eventually, scientific theories and scientific fields undergo a revolution and paradigm shift and the orthodox theories are replaced entirely by a different way of seeing the phenomenon. As Hacking suggests in his introduction to Kuhn's 4th edition, '(t)here are always discrepancies between theory and data, many of them large' (in Kuhn, 2012, p. 18). Kuhn's ideas about paradigm shift alerts us to the inevitability of anomalies building up in the system until the necessity of change becomes apparent. His theory helps us to gain understanding of the problems we are confronting through the recognition of anomalies that cannot be explained or resolved by the present paradigm. It would not be trite to suggest that the search for the appropriate resolution of humanity's pressing problems will take much trial and error before they are found. But at some point, the anomalies and distance between the theory and what those data explain is too great to overcome using the present orthodoxies, and a paradigm shift occurs. Until such a time arrives, science, in this case social science, works within the prevailing paradigm.

How might Kuhn's work apply to the problems outlined in this book? Liberal democratic society's conventional wisdom has explained social development in capitalistic terms, particularly stressing continued growth in the economy, ever rising corporate profits, and the trickle-down theory that suggests that wealth from the top will trickle down to those at the bottom of the economic ladder. The anomalies that now appear in that conventional wisdom, such as perpetual poverty, increasing concentration of wealth, and, most importantly, global warming, no longer supports the old capitalist paradigm. A new paradigm of social development is badly needed and new concepts such as the 'Green New Deal', proposed by the Democratic Party in the United States in 2018, while not sufficient in its initial form, represents an alternative to the prevailing predatory corporate capitalist model. And, while that proposal will not likely become the new operating paradigm as presently constituted, it points society in a much different direction than its present trajectory and paves the way for a new paradigm to emerge.

Kuhn's work alerts us to the inevitable need for change and a psychological process for making the transition. The world-system crisis with which we are confronted has presented current society with numerous crises which I have outlined and explained in Part 1 of this book. Unfortunately, there is no agreement among the politicians and leaders of the world to collectively recognize either the severity of the crises at hand or the potential alterations to the system that are required to overcome the situation. While some begrudgingly recognize some of the consequences produced by these catastrophes, many of our leaders are not sufficiently sensitive to the severity of the crises and the urgent need to transition to a different paradigm in order to address them adequately. Many politicians and leaders believe that retreating to an earlier period in the system is what is required, not complete system change. Kuhn's work demonstrates the total inadequacy of that proposition. The first step in achieving the goal of world-system change is to acknowledge that the anomalies in the present system are real and then to determine precisely their nature. There is no doubt that the world understands that the present liberal democratic and capitalist system is in peril – if not consciously, then certainly intuitively.

In reviewing Kuhn's work, we can tentatively describe how systems, whether physical or social, develop into a crisis. Constructing a new social order demands a world-system revolution that replaces old structures with new and appropriate constructions to deal with the crisis in the present system. Kuhn's theory on anomalies and structural crisis leading to a process of transformation can be considered to be an analogy for the socio-economic crisis that the world faces today as it struggles with futile attempts to adapt present structures to the changing conditions brought about by new technologies, global warming, and the inadequacies of the wealth distribution system inherent in the present paradigm. Does Kuhn give us any understanding of that process? It would appear that the process of transformation is messy and a matter of muddling through with many trials and errors made. It also anticipates the resistance to change that is mounted by those with a stake in the current system. Although I have made the paradigm breakdown and shift sound quite benign, it is far from that. I suspect than many hard feelings have been created over the debate about what constitutes truth in an academic discipline and what needs to be exposed as false. Much will depend on society's ability to engage in open dialogue and not continue on the path of isolating tribalism.

In addition to the physical and biological environment, humans live with one another, often in hugely populated cities, and are influenced considerably by social organization. Our behaviour is constantly influenced by the social structure in which we live. Certainly other biological organisms are organized by a social structure (Wilson, 2014), but humans have taken this phenomenon to great heights. Humans organize themselves through a local social structure, coordinated and guided by a world-system. Figure 9.3 identifies theories and critical components of theories that could be useful in providing a foundation for developing a new world-system.

Given the importance of the social structure to the functioning of human society, it is appropriate to start the investigation into world-system change, using the concept of structuralism as the basis for analysis. Structuralism has a long history

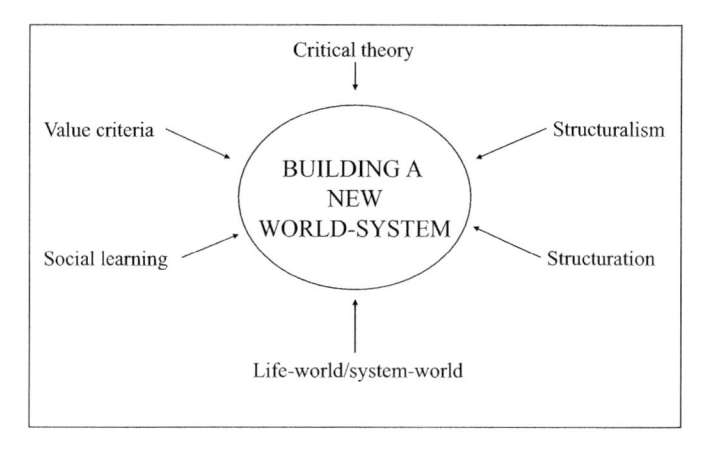

```
                    Critical theory
                         │
                         ▼
  Value criteria ╲                ╱ Structuralism
                  ╲   BUILDING A ╱
                       NEW
                   WORLD-SYSTEM
                  ╱              ╲
  Social learning ╱                ╲ Structuration
                         ▲
                         │
                  Life-world/system-world
```

FIGURE 9.3 Theories and critical components of the path to the development of a new world-system

in the social sciences as a framework for explaining many of the functions of, and impacts on, the human system. As Swingewood tells us, the:

> [S]tructuralist approach to the study of human society and culture involves the notion of wholes (a structure is not a simple aggregate of elements), the idea of transformation (structures are dynamic not static, governed by laws which determine the ways that new elements are introduced into the structure and changed) and the concept of self-regulation (the meaning of a structure is self-contained in relation to its internal laws and rules) . . . . In short, structuralism defines reality in terms of relations between elements, not in terms of things and social facts. Its basic principle is that the observable is meaningful only in so far as it can be related to an underlying structure or order.
>
> *(Swingewood, 1991, p. 296)*

Many social theorists reject the structural deterministic theory of social life. In addition to recognizing the importance of structuralism to the human system and to the analysis of the crisis in the present world-system, Giddens's notion of structuration is particularly useful because it focuses on actions by the individual in society in addition to the social organizations and institutions that permeate the system. Giddens illuminates the importance of agency in the social structure. He relies on Anderson's (Anderson, 1980) work when he defines agency as 'conscious, goal directed activity' (Giddens, 1987, p. 270). The functioning of society is not totally dominated by the social structures that exist but also by individual and collective agency. Individual agency plays an important role in making the social structure work as well as in changing it. In Giddens's scheme, the agent is not seen as a non-actor completely dominated by the institutional forces in society but actively engaged with them. Giddens gives focus to what he terms 'the purposeful, reasoning actor' (Giddens, 1987, p. 59). He goes on to explain that we need to 'treat human agents as purposive, reasoning beings, the notion of action is often understood as though it were composed of an aggregation of intentions' (Giddens, 1987, p. 60). So, in addition to the relations among the constructed elements in the social system, Giddens wants us, and rightfully so, to recognize the importance of the actor acting in, and on, the world-system. Many of our present politicians and society's oligarchs dismiss the value of individual and collective agency and attempt to suppress it. As Donald Trump has often said, he is the only one who can fix the problem. Many other authoritarian leaders from across the world may not have said that very thing, but they act and behave in similar ways.

In addition to Giddens's theory of structuration and the role of agency in social analysis, Habermas and the Frankfurt School provide a change theory in the form of 'critical theory'. The idea of critical theory depends greatly on the notion of agency. When considering world-system change from the theoretical point of view, the ideas contained in critical theory can play an extremely influential role in directing social change.

Habermas (1989) and the Frankfurt School would reinforce the role of agency provided by Giddens in the change process. Historically, '(c)ritical theories aim at emancipation and enlightenment, at making agents aware of hidden coercion, thereby freeing them from that coercion and putting them in a position to determine where their true interests lie' (Geuss, 1987, p. 64). Critical theories are self-referential and introspective; they are intended to expose the oppressive underbelly of society and focus on the coercion of the masses by powerful institutional forces, particularly the mass media and the financial corporate complex. Critical theories focus on the balance between the life-world (culture forces that support traditions, individual personality, and life meaning) and system-forces (forces that are designed to promote and protect corporate and institutional society). Figure 9.4 graphically identifies the basic principles of these forces.

The life-world can be understood most clearly through Maslow's theory of human motivation (Maslow, 1943) and his division of human needs into lower and higher order. As far back as 1943, Abraham Maslow hypothesized that humans possess at their core basic fundamental needs that require continued satisfaction, and human activity is motivated to achieve those ends. He suggested that the basic needs were ordered from lower to higher. Physiological needs are the most fundamental. The satisfaction of the physiological needs is required to sustain life. Basic nutrients, including those found in food, water, and air, constitute the basic elements of physiological needs. Fighting the COVID-19 virus is certainly a physiological need of humans. The human organism seeks physiological homeostasis at the most rudimentary level. It should be pointed out that no individual achieves 100% of each need before moving on to attain some portion of the other needs higher up on Maslow's scale. When the physiological needs are basically satisfied, humans seek to achieve their need for safety. In modern society, safety needs come

Theories of change 159

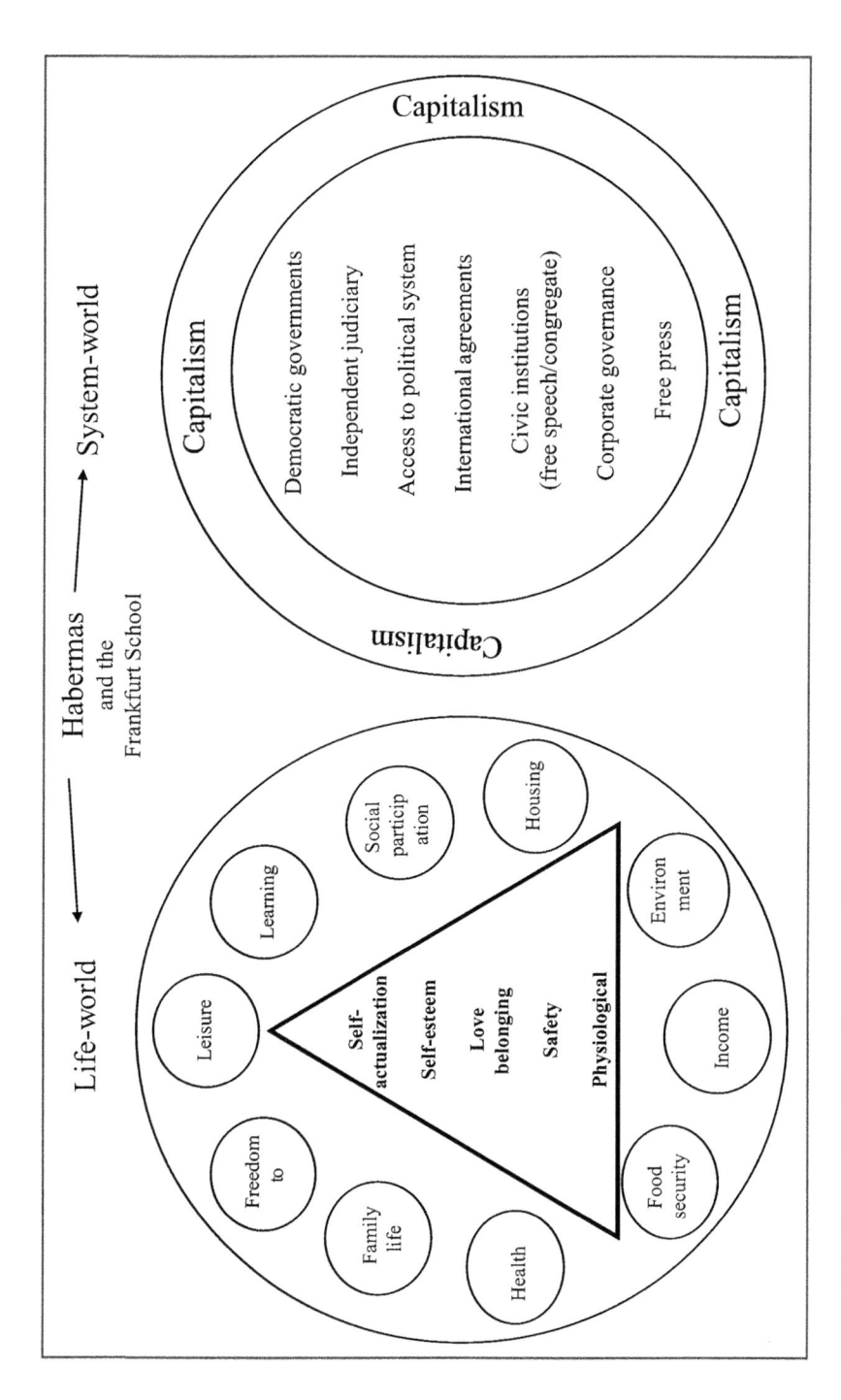

FIGURE 9.4 Forces of the life- and system-worlds

in a variety of forms. Human societies establish rules and punishments to protect their citizens from physical harm. On the international front, civilization has established an elaborate set of treaties and organizations to mitigate war or disputes that threaten our way of life or existence. Even in families, our safety is not always assured. Physical violence and abuse are of great concern in modern society especially among children and women, so society establishes policies and programs to address these issues. Inability to achieve any of the lower order needs can lead to death or pathological behaviour.

Next on Maslow's hierarchy of needs is the need for love and belonging. Maslow suggests that once the basic physiological and safety needs are met to a sufficient level, humans focus on their need for love and belonging. Although this need may incorporate romantic love, it also includes the need for family, close relationships with others, and for individual and social intimacy. Not being able to satisfy this need often leads to maladjustment in social and psychological life. Clearly, the need for belonging and love takes us out of the more basic biological need category and toward the higher order psychological sphere of human life. Belonging to an identity group is critical to the psychological health of the individual. This fact should not be forgotten when we consider the nationalism and tribalism that seems to have permeated our society today. Belonging to the tribe may be a substitute for satisfying this need often found in the workplace, which for many has been fragmented or lost completely in modern industrial society. When the workplace is no longer there, many of the needs that were satisfied in that venue are often lost along with the job. What may be clear to numerous observers is that for many people the present world-system is not able to satisfy the need for belonging and love. The ability of the present system to adequately address the remaining higher order needs may also be dubious.

The fourth rung on Maslow's hierarchy is the achievement of self-esteem. Humans routinely evaluate their life progression and aspire to see themselves in a good light in comparison to others around them. This need may provide the competitive spirit we observe in everyday life in modern human society – competition within one's self and with others. Recreation and sport have often provided access to accomplishing this need in addition to the workplace. Self-esteem provides psychological health for each individual in society. Seeking esteem by abandoning one political structure and supporting another may result if one does not attain esteem in the established social system. The unsuccessful seeking of esteem provides a possible explanation for the abandonment of liberal democracy and for the rise in nationalism and populism that we witness today. As Maslow himself stated, '(s)atisfaction of the self-esteem need leads to feelings of self-confidence, worth, strength, capability and adequacy of being useful and necessary in the world. But thwarting of these needs produces feelings of inferiority, of weakness and of helplessness' (Maslow, 1943, p. 382). Advocating for change to the social architecture to address those feelings may be a highly motivating factor for the rise in right-wing populism and nationalism that we see today.

Finally, we come to Maslow's human need for self-actualization. Self-actualization is a state that one achieves as they become the person they strive to be. Acquiring a special skill or talent may constitute such a state. Self-actualization will mean different things to different people. It is a very personalized quality. In Maslow's own words,

> (e)ven if all these needs are satisfied, we may still often (if not always) expect that a new discontent and restlessness will soon develop, unless the individual is doing what he (sic) is fitted for. A musician must make music, an artist must paint, a poet must write, if he (sic) is to be ultimately happy. What a man (sic) can be, he (sic) must be. This need we may call self-actualization.
> *(Maslow, 1943, p. 382)*

Lower order needs are deficit driven. The higher order needs are growth oriented and not deficit driven. In the era of economic transition from a labour-based manufacturing economy to a highly digital technological world, many lower and middle class people have lost access to meaningful work in the market economy and hence have lost their path to self-esteem and self-actualization.

In addition to Maslow's theory of human motivation outlined in the triangle in Figure 9.4, I have surrounded the triangle with bubbles that may operationalize Maslow's more abstract notion, at least for those living in the western nations. Certainly, these items do not exhaust the list of potential needs in each of Maslow's categories, but they provide a substantive flavour to his more abstract notion. Inspiration for this part of the life-world diagram is taken from Human Resource Canada's 'Wellbeing Diagram' (www.hrdc.gc.ca).

The circle on the right side of Figure 9.4 lays out the foundational characteristics of capitalist society. The capitalist system provides the fundamental basis on which much of life's relationships are transacted. The system-world which is based in capitalism organizes the rules on which many of our institutions are founded. Democratic societies boast that they operate by the rule of law arbitrated by an independent judiciary. An independent press is also a feature of the democratic society. The system-world also allows for freedom of speech and the congregating of people to demonstrate their views. It also encourages the development of civil institutions for carrying out much of society's business. The corporate sector is also subjected to operational rules both nationally and internationally by the system-world. From time to time, and this may be one of those times, the relationship between the life- and system-world breaks down or becomes extremely unbalanced. Theories such as the critical theories are developed to analyze the breakdown and to challenge the system at these perilous times.

Critical theories possess a normative value unlike some of the other theories presented in this chapter. Critical theories are prescriptive both in aim and in process in that they take for granted there is a social crisis and then attempt to explain the problem and how to overcome it. So they provide more than a framework for analysis and explanation in that they predict a course of action based on a particular understanding of the social system and the workings of the institutions within it.

The concepts provided by the idea of critical theory offer a basic framework or rationale for critiquing the present world-system and for providing processes for creating remedies for its transformation.

Economic theories provide a slightly different theoretical point of view to the world-system crisis and transformation discussion. Economics is a part of social science, 'and there is no justification for keeping economics as insulated from other social sciences as many economists seem to presume both possible and desirable' (Giddens, 1987, p. 259).

In any discussion on the role of economics in social change, we need to recognize the contributions of such notable thinkers as Adam Smith (1771), John Maynard Keynes (1936), and Karl Marx (1867), among others. These notable individuals represent contrasting ideas about how economics influences social organization and human relations. What is important to note is that their works are not just about economics but also lay out ideological concepts and notions of how society works best.

Adam Smith explained how a capitalist system works to attain economic efficiency in human relations. Smith's ideas formed the basis of the capitalist system that we revere today. Karl Marx reacted to Smith's notion and maintained that raw capitalism would eventually exploit the worker in favour of the owners of capital. Capitalism, in his view, would simply make the rich wealthier and the poor poorer. Capitalism produces winners and losers, with the winners becoming fewer but wealthier and the losers gaining in numbers and becoming poorer as the system matures. We must consider Marx to be somewhat prophetic given that OXFAM has recently announced that 42 people in the world now own as much wealth as the bottom 50% of the world's total population – that's 3.6 billion people.

John Maynard Keynes provided an adjunct theory to the basic notion of capitalism. He legitimized the intervention of governments into the economic system if only to stimulate growth when it appeared to be headed for recession. Keynes came along at a critical point in the history of world capitalism, the Depression of the late 1920s and early 1930s. He argued that governments needed to generate demand by creating 'make work' programs through infrastructure projects and other social programs during times of recession. He theorized that by injecting government money into the faltering economy by way of creating public infrastructure projects, private investment would be stimulated. Keynes's theory is given considerable credit for shortening the duration of the Depression of the 1920s and 1930s. During the advent of capitalism, the philosophies of Smith and Marx were the major protagonists. Keynes came along at the time of crisis and legitimized intervention in the capitalist system in the form of fiscal and monetary policy.

It is wise to remember that Adam Smith and Karl Marx were addressing economic systems that had yet to mature fully. Since that time, they have fully matured, and some would argue that neither of their theories is appropriate for addressing today's issues, yet we still discuss them as if they do.

Is there a set of ideas that stand the test of time yet still relevant for today's issues and problems? Perhaps Polanyi's *The Great Transformation* will provide a more useful framework for this discussion. As the introduction to Polanyi's work provided by Fred Blok suggests:

> Although few books these days have a shelf life of more than a few months or years, after more than a half century *The Great Transformation* remains fresh in many ways. Indeed, it is indispensable for understanding the dilemmas facing global society at the beginning of the 21st century.
>
> *(Polanyi, 1944, p. xviii)*

Polanyi was one of the first economists to critique the general consensus that self-regulating markets would operate in a natural fashion and seek equilibrium. He was an advocate of government regulation to guard against excesses in the free market that would damage many people, particularly the poor. It is worth citing Polanyi himself as he describes the idea of the self-regulating market and the focus on growth:

> The assertion appears extreme if not shocking in its crass materialism. . . . Nineteenth-century civilization alone was economic in a different and distinctive sense, for it chose to base itself on a motive only rarely acknowledged as valid in the history of human societies, and certainly never before raised to the level of a justification of action and behavior in everyday life, namely, gain. The self-regulating market system was uniquely derived from this principle.
>
> *(Polanyi, 1944, p. 31)*

In some ways, Polanyi anticipated the 2008 financial crisis. He was one of the first economists to challenge the idea of trickle-down economics that asserts that if the wealthy do well, so will those below them on the economic ladder. He was a leader in arguing that there is a relationship between ideology and self-interest, that is, the idea that concentration of wealth tends to lead to plutocracy. Polanyi challenged the notion that the world could be a better place if it would only accept the neoliberal idea of the self-regulating market and reduce governmental regulation to a bare minimum or stop intervening altogether. This idea elevates the market to the predominant social institution in society rather than an embedded organization within the social structure. He saw democracy becoming embedded in capitalism and thought it should be the reverse.

From the 1950s to the 1980s, the economic system became more compressed than it was before World War I (Krugman, 2007). The middle class grew quickly and increased dramatically during that time. Government regulation of the markets, including such statutes as the Glass–Steagall Act instituted in the 1930s, constrained the activity of financial institutions, separating commercial and investment banking until 1999 when it was repealed. We are now heading into what Nobel Prize winning economist Paul Krugman (2007) refers to as a second Gilded Age where there is extreme separation between the rich and the rest of us. Polanyi's work can be instrumental in providing a framework for analyzing the conditions leading to the transition we are confronting at present. Further, his ideas may also act as a destination beacon as we move into the future.

This chapter has presented theories that help explain in a focused way the basis on which the social system has been formed and operates, and theories that may provide a framework for change as we transition from a predatory corporate capitalist system to a more sustainable world-system. I have presented the bare rudiments of each theory and explained them in context of world-system change. I invite the reader interested in these subjects to immerse themselves in the literature that could provide more detail on each of these subjects.

To sum up, the theories presented in this chapter explain several aspects of world-system change. Emotional intelligence, in addition to reason are important in their different ways to the decision making process. In addition to describing the three basic modes of reason – instrumental, aesthetic, and moral – it lays out the basis on which decisions should be taken. This section of the chapter suggests that instrumental and aesthetic reason must be enveloped by moral reason given the conditions in the social and environmental world today. The discussion of emotional intelligence and the theory of reason adds a foundational element to the social-change process.

The discussion on structuralism, structuration, critical theory, life-world/system-world balance are included here to provide the reader with the basic ways social science views how the world operates. Structuralism places emphasis on the institutions in society, while structuration introduces the role of people (agency) in those institutions and in orchestrating society. Critical theory, on the other hand, provides an approach to world-system change that places emphasis on the emancipation of the masses in society and empowers the goal of world-system change. It is a critique of the power institutions that have gained prominence over time and now dominate society. As part of the goal of world-system change, the rebalancing of the life-world with the system-world is thought to be critical to social change.

Selye's GAS and Kuhn's paradigm-change theory offer an explanation of how world-systems, either biological or social, function under stress and change, or don't and become dysfunctional or die out completely. Maslow's notion of human motivation reminds the reader that the basic needs of humanity must be satisfied for the species to continue and, therefore, need to be kept central in the social change discussion. These theories establish the reality of the dynamism of the social, cultural, and ecology of the planet. They provide a lens through which we can examine and analyze the present world difficulties. Kuhn's and Selye's theories explain how organisms and systems change and what happens when adaptation to a new environment is not transitioned successfully. Maslow adds the human motivational dimension to the discussion.

Systems theory, chaos, and complexity theory have a paramount role to play in this discussion and in world-system change generally. The social and physical world is a complex system that acts much like entropy, the second law of thermodynamics, in that it is continually moving from order to chaos and, as a result, must be understood as complex and an interconnected system. Most fundamentally, chaos/complexity theory alerts us to the fact that the world is constantly changing, and our social structure must change and adapt to those changes. The ideas of contingency, accidental, or conjuncture events provide the unpredicted actions that spark change to some of the variables in the system. Kuhn's notion of anomalies building up in the system making it ineffective over time and in need of significant change is apropos to the discussion of world-system change. Finally, a small but important theoretical discussion in this chapter is the use of probability theory in social analysis and prediction. Certainly, chaos and complexity theory understand the necessity and utility of this approach to understanding. Chaos and complexity theory appreciate that the destination of the system is indeterminate, therefore not predictable.

There is always a major gap between understanding what needs to be done about a social problem and how that issue can be addressed and resolved. The difficulty becomes even more pronounced as the issue and the context in which it resides becomes larger and more complex. Thinking and working at the world-system level presents enormous challenges. Regardless of the size and magnitude of the task, a project of this nature requires a framework on which a process can be built and implemented. This chapter has provided some theories that can help construct a way of viewing the present social situation and useful for constructing frameworks for change. The next chapter continues the discussion and introduces methodologies, frameworks, and processes for change.

Bibliography

Anderson, P. 1980. *Arguments Within English Marxism*. London: Verso.
Bertalanffy, L. V. 1956. *General Systems Theory Revised Edition*. New York: George Braziller.
Boulding, K. 1956, April. General Systems Theory: The Skeleton of Science. *Management Science* 2, no. 3: 197–208.
Byrne, D. 1998. *Complexity Theory and the Social Sciences: An Introduction [Kobo version]*. Kobo.com.
Galbraith, J. K. 1967. *The New Industrial State*. Princeton: Princeton University Press.
Galbraith, J. K. 1984. *The Affluent Society*. Boston: Houghton Mufflin.
Geuss, R. 1987. *The Idea of Critical Theory: Habermas and the Frankfurt School*. London: Cambridge University Press.
Giddens, A. 1987. *Social Theory and Modern Sociology*. Cambridge, UK: Polity Press.
Habermas, J. 1989. *The Theory of Communicative Action*. Oxford: Polity Press.
Hobbes, T. 2012. *Leviathan*. New York: Start Publishing.
Jeffko, W. 2018. *Contemporary Ethical Issues 4th ed. [Kobo version]*. Kobo.com.
Keynes, J. M. 1936. *The General Theory of Employment, Interest, and Money*. London: Palgrave MacMillan.
Krugman, P. 2007. *The Conscience of a Liberal [Kobo version]*. Kobo.com.
Kuhn, T. S. 2012. *The Structure of Scientific Revolutions 4th ed. [Kobo version]*. Kobo.com.
Locke, J. 1689. *Two Treaties of Government*. London: Mobile Reference.
Marx, K. 1867. *Das Kapital. Kritik der politischen Oekonomie*. Frankfurt: Verlag von Otto Meisner.
Maslow, A. 1943, July. A Theory of Human Motivation. *Psychological Review* 370–396.

Pentland, A. 2014. The Rational Individual. In *The Idea Must Die: Scientific Theories That Are Blocking Progress*, edited by J. Brokman, 597–601 [Kobo version]. Kobo.com.

Polanyi, K. 1944. *The Great Transformation: The Political and Economic Origins of Our Time*. Boston: Beacon Press.

Rousseau, J. J. 1988. *The Social Contract [Kobo version]*. Kobo.com.

Selye, H. 1956. *The Stress of Life*. New York: McGraw Hill.

Skyttner, L. 2001. *General Systems Theory Ideas and Applications*. Singapore: World Scientific Publishing Co. Pte. Ltd.

Smith, A. 1771. *An Inquiry into the Nature and Causes of the Wealth of Nations*. London UK: W. Strahan and T. Cadell.

Swingewood, A. 1991. *A Short History of Sociological Thought, 2nd ed*. London: MacMillan.

Waldrop, M. M. 1992. *Complexity: The Emerging Science at the Edge of Order and Chaos [Kobo version]*. Kobo.com.

Wallerstein, I. 2004. *World-Systems Analysis: An Introduction [Kobo version]*. Kobo.com.

Wilson, E. O. 2014. *The Meaning of Human Existence*. New York: W. W. Norton & Company.

10
PLANNING FOR CHANGE

> [W]hat might a twenty-first century version of a mixed economy look like? In this climate, policy ideas need to pass several tests. They need to be sensible in their own right, produce real benefits for broad constituencies, contain the power of elites (especially financial ones), revive the power of the common people, rebuild democracy, expand the realm of 'decommodified' forms of social income that are not dependent on market wages, and teach durable lessons about the failures of laissez-faire. They need to rebuild an ideology of solidarity and common purpose using national democracy, and reclaim space from the global market. And they need to restore confidence in public solutions and public institutions . . . A new version of a social economy needs to be even better defended institutionally and politically than the last one was.
>
> (Kuttner, 2018, p. 487)

Kuttner's assessment of what might constitute the basics of a new world-system is a good starting point for discussing what should come next. Although it is intriguing and useful to outline the possible structure of a new world order, it is, perhaps, more important to determine the strategies for achieving that state. This chapter provides methodologies, processes, strategies, and tactics for tackling the question of how society can proceed to changing the predatory corporate capitalist system to something more egalitarian, pluralistic, democratic, and environmentally sustainable. The methodologies, processes, strategies, and tactics outlined in the chapter rely on the theoretical foundations provided in the preceding chapter.

This chapter shifts the discussion from examining and explaining the social and environmental conditions driving the need for world-system transformation and the theories helpful in explaining social change, to designing appropriate actions for implementing the transition. The range of options available stretch from relying on the invisible hand of the market to solving the crisis, and hoping that it works, to proactive interventions in the social and environmental systems through analysis and planning. Given there is no evidence for an invisible hand guiding human development, as some would have us believe, human intervention in both the environmental and social systems is the most appropriate path to follow. This chapter examines the rationale for planning world-system change in addition to processes and strategies for implementation. Transactive Planning that embraces the value of engaging people in the planning process underpins the approach to system change advocated here.

Part of the rational thought process in connecting understanding to action is the idea of planning for the future. All of us engage in this process on an individual basis as well as collectively. Humans may be one of the few, if not only, species that contemplates the future and arranges actions to arrive at a predetermined goal. At times, this process is subconscious and at other times very much in the conscious foreground. When describing planning, Carroll tells us,

> There's a lot going on beneath the deceptively simple idea of 'making Plans'. We have to have the ability to conceive of times in the future, not merely the present moment. We need to be able to represent the actions of both ourselves and the rest of the world in our mental pictures. We must reliably predict future actions and their likely responses. Finally, we must be able to do this for multiple scenarios simultaneously, and eventually compare and choose between them. The ability to plan ahead seems so basic that we take it for granted, but it's quite a marvelous capacity of the mind.
> *(Carroll, 2017, p. 350)*

Carroll's description and notion of planning is based on creating a foundation of empirical data. It is not planning to simply call on the spirits and ask them what is in store for humanity in the future and then create a plan that reflects those ideas. Planning future scenarios needs to have a strong relation to the present context, and the plans developed must incorporate and respond to the interpretation of empirical data. Plans should be consistent and relevant as much as possible to the data collected and analyzed. This suggests that great care needs to be taken in determining what data are necessary and useful to the matter under review. The determination of the data to be collected is part logic and part political, so it can be fraught with danger. Planning is the process of turning the 'ought' into the 'is' or, to put it in more academic terms, taking action through the interpretation of an empirical analysis of data in light of selected goals and objectives for the future. As Carroll notes, 'science describes the world, but not what we're going to do about it' (Carroll, 2017, p. 430). Further, the scientific method has nothing to do with the decisions about what to study. That is a political question, and data selection must be clearly described and open to scrutiny by all those the plan will eventually affect. The remainder of this chapter is devoted to fleshing out the methodology of planning and the values on which the planning activity is based.

Historically, humans have continually struggled to create a society that provides opportunity, dignity, and social justice for all. Wharf and McKenzie (1998) provide the basis for planning and system change when considering those factors. They tell us that a deep understanding of the unacceptable condition is often the focus of planning. The concern generating a desire for change comes from a dialogue among large numbers of those affected by the issue. The build-up of oppression in society needs to be exposed and addressed. Most people come to realize the consequences to the human condition if the undesirable state is not altered and continues to build. At a critical point they decide to do something about it. Although data-based transformation has not always been available in historical times, today's oppression can be quantified as well as felt emotionally. Perhaps the most abstract but relevant antecedent to sociocultural evolution exists when the social system no longer seems adequate to address the basic needs of individuals or when large groups of people become marginalized in society and find it increasingly difficult to accept their diminished lifestyle or the increasing social injustice perpetuated by the system. Realizing that something is wrong in the system occurs when people begin to experience social and cultural breakdown. Social cohesion unravels and eventually collapses. Individuals step back from their day-to-day activities and discover what they thought they had established as a viable structure built on a core set of values is no longer operative or adequate for conducting everyday life. When this condition creeps into society unnoticed, it is often described as democratic drift or political decay. However, when there is breakdown in the social system, there is often motivation and opportunity for sociocultural change. Breakdown in society may lead to breakthrough in terms of creating a new social architecture and contract. The recognition that the social system is in a state of chaos is a prerequisite to engaging in a transactive planning process that is dedicated to large change in the social system.

As part of establishing a robust dialogical structure and eventual plan of action, analysis of the extent of environmental degradation and social injustice in the system can be gleaned by the examination of rich data bases that are supplied by governmental and non-governmental organizations. Data collected by both government and independent bodies is extremely useful and important to this process. Modern society will need to rely more and more on evidence-based decision making in the social policy planning process as the quantity and value of that research becomes increasingly understood and appreciated. A combined balance of empathy and empirical evidence needs to be judiciously employed in the social policy formation process. Incorporating these factors in decision making is necessary if rational life-sustaining policies are to guide us into the future. And while politics will always envelope policy decision making, accountability can be provided through research and analysis of rich data bases that inject non-partisan information into the decision making system.

Historically social change has been opportunity driven. What is critical to making successful changes in the social architecture is the understanding of how new developments and new ways of seeing and thinking about the world can lead to a better and more sustainable reality. This cognitive state is analogous to Kuhn's notion of paradigm shift in science. The only way to achieve this understanding is to educate the public on the importance of achieving the proper balance between the life-world and system-world. Social dialogue that leads to defining the balance point on that continuum is the beginning stage of developing a new social contract. The relationship between these two forces is completely unbalanced at the present time. One could argue that this state is a result of too much faith given to the free market forces in the capitalist economy. There has been much false consciousness glorifying the market built into the present public discourse and this false consciousness will need to be addressed before we can determine what is the correct course for today's society. We constantly make assumptions about truth and falsity, and the overwhelming reliance on the classic economic system may be a false assumption in today's world. The public needs to take control of the unfettered market and make the economic system work for everyone, not just a few.

It must be stressed that no social action can be successful without first implementing a public discussion (education program) focused on clear and truthful dialogue on the problems confronting society and the changes essential to rectifying those difficulties. Habermas' notion of life-world and system-world can provide a framework for that discussion. We need to put traditional assumptions aside and rely on the facts uncovered by empirical evidence. Public understanding of the core issues and the rationale for change is a priority in the transformation process. No transformation will be successful without the absolute support of the general population, and a well-developed public education process and dialogue is an essential part of getting to that point. Education in this context is meant to be experiential, not the inculcation of the masses to the ideas and goals of the powerful members of society.

Today's transition needs to be based on achieving *planetary sustainability* and *social justice*. To achieve these goals, there are some fundamental principles and prerequisites that need to be recognized prior to the action-planning stage of transactive planning. Transactive planning relies on two fundamental principles. First, it hinges on data analysis to explain the anomalies in the system and then the patterns and forms that social change can take based on that analysis in dialogue with those who will be affected by the eventual plan. Important to this analysis is the discovery of the context and trends that lead to change. Uncovering the threshold points at which motivation stimulates action is an important consideration when exploring social change through planning. It is important to recognize that constant feedback is a critical criterion for explaining the changes taking place in the system as the plan is being implemented. People involved in and affected by proposed changes need constant feedback regarding the process and the stage it is progressing through at any point in time. Second, two-way communication is key. How populations respond to a policy, or the lack thereof, is a critical dynamic in the progression of society. Constant monitoring and evaluation of the actions taken, or policies implemented, must be serious components in the planning process. Actions taken based on any plan will need adjustment and refinement as experience dictates, and policy implementers must not be reluctant to adjust the plan when required. Often, the requirement to change direction comes from those intimately exposed to the changes taking place. They are often the first to feel the impact of change, and their responses must guide further development of the system.

The amount of impact on deeply held values demanded by change is an influencing factor in the social transformation process. Deeply held values constrain the amount of change a paradigm shift may produce or the speed with which a plan can be implemented. We are emotionally attached to our way of life, and alteration to that lifestyle is often hard to implement or accept. Additionally, the amount of time a shift in the social system may take can be lengthy for some and too quick for others. Society's ability to allow enough time for affected populations to adjust to a changed lifestyle a new social contract may demand is critical. We have often witnessed the difficulty that immigrant families encounter when they are immersed in a new society whose social system is constructed on a very different set of values than those of the old home country. There is often great conflict between the generations as to which set of values will dominate individual and social life. On the other hand, change must occur in an appropriate timeframe and address the issue in an acceptable time interval. Delay has become a tactic on the part of some who would not want to see the plan implemented or the status quo changed.

A paradigm shift is not just a social phenomenon but a psychological one as well and supports need to be put in place to assist individual members of society to make the transition. The proposed changes to a society need to speak to the problems that are affecting most citizens, and the proposed remedial measures to be taken must be seen to have the ability to enhance the social cohesion of society. If there are clear winners but also clear losers, the proposed change is likely to be resisted. Resistance must not be ignored but examined by those implementing the project, and plans put in place to ameliorate the negative effects.

What is most clear about the present period is that there is no natural force or invisible hand that will guide us benignly into the future. Consequently, the transition will be stressful for many of our fellow citizens. Regardless of the potential for disruption, we must anticipate the future and plan accordingly for it. Although anticipating the future is fraught with danger, there can be some clear indicators in the social and physical environment on which to proceed. The scientific data available on the concerns expressed in Part 1 of this book and the current anxiety among the population requires us to act. And, although data do not make decisions about future directions of society, it can aid decision makers greatly in making quality decisions.

Any successful plan will need to be imbedded in a massive education and communications program. Successful change to the social architecture at the world level will require an enormous education and consultation campaign to determine, and then explain, what is in society's and the planet's best interest. The first order of business is to engender a recognition by society that the present world-system is no longer working and, in fact, is now in a chaotic state. Although we may not agree with the authoritarian approach to change that is taking place today, what is clear is that a good portion of society understands that the status quo is no longer tenable. It is imperative to proceed on a basic understanding by the mass of society that the present system has broken down and requires transformation. It is clear to me that society for the most part has arrived at that conclusion even though it is a bit chaotic in its present expression.

Social activists eager for change often attempt to circumvent the necessary first steps of community education and organization and move directly to taking immediate substantive actions to solve the problems they deem important. Leaders of the process must not move too fast and get out ahead of those they lead. Those who form the vanguard of problem recognition must constrain their eagerness for instant action. A critical part of any social action is spending time and energy building mass support through identifying the issues that resonate with the public. Not until a shared understanding of the matters to be addressed and a program for addressing them is reached, can action be taken. The experiential education of the public to their oppression is a necessary first step without which the project is doomed to fail. This is evidenced by the Tea Party and the followers of Donald Trump, as well as the Brexit advocates in the United Kingdom. Although we may not agree with the solutions these groups advocate, their vociferous discontent is a clear demonstration that all is not well in society. They have accurately determined that society is not working for them, but they have not, I would argue, landed on the actual cause for their dissatisfaction or engaged the appropriate remedial actions for rectifying their distress. The identification of the cause of their discontent is left unarticulated. Their position is built on assumptions and an emotional reaction to their plight. They have been unable to transition out of the emotional stage of their discontent and enter the analytical phase before acting. The process has been truncated. It is impossible to determine the proper course of action if the problem to be addressed is wrongly identified in the first instance or stated in the form of a solution. When that happens, what you are likely to get is some form of 'Make America Great Again'. The public has fallen back on emotionality and ideology rather than assessing the data to correctly identify the problem and then determine the best course to correct it. Their unrest should not go by unnoticed, however. The emergence of these groups and their actions is clear evidence that there has not been meaningful research or public dialogue on these matters before jumping to assumed solutions (build the wall). What's more, many people have a deep mistrust of government that must be overcome if any progress is to be made on macro change to the world-system. Attention to increasing social solidarity needs to be a major effort on the part of those desiring change and a public discussion on the inability of the present world-system to satisfy today's needs is a fitting place to start. So, there needs to be some agreement throughout society that there is a problem to be rectified and its details articulated.

Problem identification and clarification in the modern world is based on the notion that all human problems are political and not simply technological. As we are often reminded, our global problems are not technological ones even though advancing technology may eventually play a large role in resolving some of these issues or in making things worse if not directed appropriately. Even so, all problems are political in the final analysis and contain conflicts over values, interests, and power. In fact, it can be argued that the problems from which our present system suffers are moral issues, and neither economic nor technological solutions will address them adequately, although economics and technology will have a role to play in their eventual resolution. Only through planning that relies heavily on collective problem identification, then on vision creation, followed by collective action by all citizens in society, will humanity's present condition be addressed satisfactorily. All solutions chosen for implementation will need the full support and buy-in of the general public if there is any chance of successfully addressing the deleterious conditions that exist in society today. No doubt this prescription is a tall order to realize fully in today's social and political climate, but nothing short of this kind of social dialogue will set civilization on a sustainable path.

The process of problem identification needs to be constructed on two main beliefs about present society. First, the current social organizing strategy – predatory corporate capitalism and the economic growth model – in its present form is now exhausted and no longer adequate to serve the true interest of society. Second is the realization that the present version of capitalism has devolved from the variety envisioned by Adam Smith to predatory corporate capitalism that is clearly slanted in favour of those with money to gain more of it and for those at the bottom to continue in subjugation and servitude. Chrystia Freeland (2012), in her book *Plutocrats*, identifies a three-stage process to engineer social change. First, the wealthy gain complete control of the world. They become actively engaged in making government policy that favours them and subjugates everyone else. Second, the mass of society determines that the status quo is no longer tolerable and identifies alternative methods for organizing society's affairs. Finally, society legitimizes a new paradigm that can be implemented to meet the crisis (Freeland, 2012). It would appear we are heavily engaged in Freeland's process.

Without doubt, the first stage of Freeland's schema is evident today. We may be in the middle of the second phase and are just now beginning to see some ideas emerging regarding alternatives to the present condition. Trumpism may be one of those ideas but the Green New Deal may be a more promising one. Public discussion on these competing ideas within society needs to be given more attention than they are getting now. At present the differing tribes in society are attempting to apply their ideas with force and without fully airing their virtues or limitations in the public domain and in a civil manner. There is no public dialogue only threats and a lot of mean-spirited quarrelling at the moment. Forcing ideas on others seems to be the method of choice to replace public dialogue, but that tactic will get us nowhere fast.

Truncating the dialogical and educational part of the planning process to get to the proposed remedies immediately (maybe not the right ones) will only cause conflict and division among the public much as we witness today. The problems that are attempting to be rectified, and the solutions chosen to rectify them, are someone's idea of what is important to address but those efforts do not have the agreement or support of the majority of the public. The type of education envisaged here must be bottom-up social learning that requires dialogue among the public and not top-down slick marketing or coercion by those in charge of the present system. The sociocultural evolutionary change that is required to address the present critical social and environmental crisis will entail a sound transactional social planning format that is rational and involves an extensive social learning component. It will also require the establishment of a social marketing thrust and the development of strong international mechanisms to facilitate the process.

When I speak of education, I speak of a dialogue among people to collectively determine what is in the best interest of humanity and the planet and not a top-down approach similar to what goes for education today. Freire (2007) makes a distinction between the 'banking' and 'problem-solving' approach to education. Fundamentally, the banking system allows the participant in the discussion to enter the expert's consciousness, whereas what is called for here is for members of the public to construct and enter their own consciousness in dialogue with each other. The expert or planning agent must enter the consciousness of the general population. It is truly a collective development of awareness and conscientization.

The type of education suggested here focuses on the emancipation of the population from the critical life issues oppressing them, not on gaining skills for employment or knowledge for knowledge's sake (banking system education), although there is a place for that type of learning (understanding the technical problems of climate change, etc.) but not the primary method employed in the social change process. The educational effort envisioned here would focus on collectively determining what is negatively affecting life such as climate change and the unequal distribution of wealth, and not continue to teach the prevailing myths and falsities such as the primacy of classical economics for curing all individual and social ills that permeate our discourse today. It is important to understand that our own interests would be enhanced by cultivating the wellbeing of others and increasing social justice while paying less attention to market forces for organizing life. By strengthening collective interests, we advance individualism. It is the collective that allows for individualism to flourish. Strong collectives maintain and enhance individualism, and strong individuals understand the need for strong collectives.

The alternative to the market for making decisions that affect relations between and among people is transactive social planning and direct intervention by the human community in the social and environmental system. This approach intentionally arranges relationships and events in a rational and directed manner rather than leaving human affairs up to the so-called invisible hand of the market. Although most capitalist countries would lead you to believe that they operate in full unfettered market economy where the market makes all decisions about the direction and order of society, planning in all forms plays a bigger role than most people think.

Planning is a very human activity. It is a function of designing the future. It is highly values-based. Planning is carried out by governments, private business, and civic organizations. It is a part of carrying out individual daily lives whether it is a conscious act or not. Emphasizing agency (Giddens, 1987) and implementing the ideas contained in critical theory that stress communicative action throughout planning (Geuss, 1987) are considered to be fundamental to world-system change. These ideas may be best exemplified by Friedmann's notion of social mobilization (Friedmann, 1987) and Giddens's ideas of structuration. This approach to planning is very much bottom up, value laden, and highly participatory and is often referred to as transactive planning.

Transactive planning provides a structure and format for driving social and environmental change. At its most fundamental level, the stages in the planning process are, identifying and articulating the problem to be solved; envisioning a desirable future that addresses the problem to be rectified; identifying and articulating the goals and objectives necessary for achieving that vision; establishing strategies and tactics for reaching each objective leading to goal achievement; laying out an implementation strategy; and creating a process for continually evaluating and monitoring the progress of the planning activity and the outcomes it produces to be sure the actions taken directly address the goals and achieve the initial vision. The evaluation and monitoring phase of the process will also allow for adjustments to the plan to be executed along the way. A highly important but often overlooked part of the transactive planning process is the creation of an implementation plan and strategy. The greatest plan in the world needs to be executed if its contents are to become reality. As stated earlier in this chapter, simply creating a plan without a robust implementation program and a management strategy is not good enough to make the transition to a new world-system successfully. Developing an outreach strategy that involves as many people in the movement as possible is a major task and should not be overlooked. The success of the movement and the entire project is dependent on gaining enthusiasm within society for what is being proposed. It must not only appeal to logic but also engender emotional acceptance. Humanity must see the value of the path chosen and engage in the journey with enthusiasm. The legitimacy of the proposal and the transparency of its activities are crucial.

As stated earlier, the first stage of the planning process is setting out the problem to be addressed by a vision statement that is created to guide the project. It is critically important to state the problem accurately. The task is to keep the goal of the project and the means to achieving that goal straight. It is too easy to get the goals mixed up with the means for achieving them. In my travels as an academic and planning consultant, I have often heard people state, 'our problem is we need a new facility of some type', which is clearly not a problem statement. For example, I have heard municipal councils state a problem as a need for a new ice hockey arena. That is a statement of a potential solution to a problem yet to be identified but not a problem statement. After being queried, the problem was identified to be a rising level of juvenile delinquency in the town to which an arena would be one among many potential solutions. This type of mistake will start the planning process down a road no one really wants to travel. Taking action prematurely and before the problem to be addressed is clearly articulated and agreed to by all concerned parties is bound to lead to chaos and frustration, exactly what we see in society today. I often get the mental image of someone standing in front of a slot machine automatically feeding it coins, pulling the handle, and hoping to hit the jackpot, or in this case the correct problem that needs addressing. Sloppy problem identification and articulation will only lead to disaster. Much effort and focus need to be placed on understanding the locus for the anger and frustration that permeates society by assembling fact-based information that can lead to a clear statement of the problem(s) to be addressed. This method of attacking problems may prove more fruitful in the long-term than moving to solutions prematurely and from only an emotional or ideological point of view. It must be remembered that how a problem is conceived and articulated dictates the range of possible solutions available to remedy that problem. If a problem is seen as an economic growth problem, then increasing the GDP may appear to be a reasonable solution. On the other hand, if the problem is determined to be an inequality issue – a moral question – then distributing the resources available equitably could form the preferred option. How a problem is perceived and articulated is critical at the start of the process.

Part 1 of this book and subsequent discussion in Part 2 could lead to the conclusion that one of the problems needing attention is the inability of the present predatory corporate capitalist world-system to adequately address and rectify the issues outlined throughout Part 1 of this book. Of course, the initial statement of the problem would need to be detailed out and fully vetted, but once agreement on the problem is reached, a positive vision statement that addresses that problem can then be created. Goals and objectives for each of the domains outlined in Part 1 of this book that require action can be struck. Achieving these goals and objectives in turn will realize the vision. In the final stage of articulating the plan, measurable strategies and tactics designed to accomplish the goals and objectives are developed and implemented. The strategies and tactics must be designed with their potential impact on the other domains in the system in mind. Monitoring and evaluating the processes and outcomes of the plan follow. Evaluations would be conducted on the action of each part in the system with respect to the other parts and the potential overall effect on the system as a whole.

In addition to the critical vision creation stage of the transactive planning process outlined earlier, there are other stages that need attention in order to complete the process. For example, who should be involved in programmatic and alternative selection and on what basis will these issues be pursued? The same goes for carrying out an evaluation on the alternatives. Of course, there is no pat answer to these questions other than to say they will need to be negotiated. Certainly, national governments will need to be major players in the process. The path to figuring out this process will be partially addressed through the discussions later in this chapter on capacity building, networking, organizational development, and social learning.

While these technical elements of the planning process seem relatively straightforward and simple, they are imbedded in a highly participatory, complex, macro framework as outlined by Giddens, Habermas, and Friedmann and discussed in Chapter 9. Following a framework and process that is highly participatory not only leads to accomplishing the collective vision, goals, and objectives set out in the plan but also leads to social learning on the part of everyone involved. It strengthens citizenship and leads to individual and collective empowerment as well as achieving the basic goal of world-system change. It is a function of designing the future. It can build social solidarity by promoting the engagement of individuals in planning and creating a future that promises to better the lives of all citizens and not just the powerful and wealthy few.

Wharf and McKenzie (1998) argue strongly that the planning process and its outcomes must be based on a series of value criteria (refer to Figure 9.3). They suggest that value criteria should lay the foundation for planning sociocultural alteration. The approach should be explicitly critical in considering historical, cultural, political, and economic factors. People must be recognized as active agents in shaping as well as reacting to their environment. The life experience of users must be considered. I interpret these value criteria to be similar, if not exactly, to what is contained within the life-world. The balance between life- and system-world can also be employed as a criterion for evaluating the policies and actions driving change. If any action is not seen to enhance the life-world, it should be rethought. The present world-system is based on the values of capitalism which has come to favour the system-world. The new world-system would replace those values with the principles contained within the ideas of economic equality (limited inequality), social justice, civic leisure, and environmental sustainability, to name just a few.

It may seem surprising for the reader to see a case being made for a planning approach to sociocultural development. It is generally thought that change, as large as the one facing humanity today, should happen automatically and spontaneously and without need for embracing a mechanism for making that change happen. Isn't evolution supposed to happen automatically and without forethought? Isn't evolution dependent on gene mutations that fit better to the changes in the physical and social environment? Until modern humans arrived, that was the case. So, what is different with this stage of human history that changes the mechanism of development? What is different is that we need to anticipate the changes that are required for continued human existence and create a future path that is set in motion through rational actions. In cultural evolution the change mechanism is us collectively, not the isolated gene. The exceedingly fast pace of technological change and its social consequences are too rapid for the gradual change to which society has become accustomed. Gradual evolutionary change is now too slow to be of use to human society as we move forward into the future. We no longer have the luxury of adapting to a slowly changing social or physical environment. The declining health of the planet and social system is now too fast paced for that. So far humans have made some remarkable adaptations and it is now time for another one. We need to change our culture because of where our technologies have led society and in anticipation of where our new developments are taking us in the future. Fundamentally, we cannot survive either physically or socially unless we drastically change the direction of our present social structure where all wealth is amassing in the hands of the wealthy and where our physical environment is under considerable threat because of our abuse of the planet. That abuse, by the way, is a

178 Processes of change

direct result of our past successes. Because of the circumstances now facing humanity, the employment of a rational, transactional, planning framework is required to move society forward into a new social and environmental contract. As John Shaar put it some time ago:

> The future is not a result of choices among alternative paths offered by the present, but a place that is created – created first in the mind and will, created next in activity. The future is not some place we are going to, but one we are creating. The paths to it are not found but made, and the activity of making them changes both the maker and the destination.
>
> *(Shaar, 1989, p. 321)*

Planning is not just a blueprint for what needs doing but also a framework for bringing people together to decide what needs to be done and then determining how to go about doing it. As Shaar's quote points out, it is as much about self and community-education as it is about creating a vision for the future and then goals, objectives, and strategies for implementation, although it is about that too. Shaar sets the tone for proceeding to plan the future. Figure 10.1 lays out in broad strokes the major elements that can provide a framework for planning the project of world-system change.

Figure 10.1 describes in broad strokes the fundamental elements that could contribute to the world-system change process. The outer ring encompasses

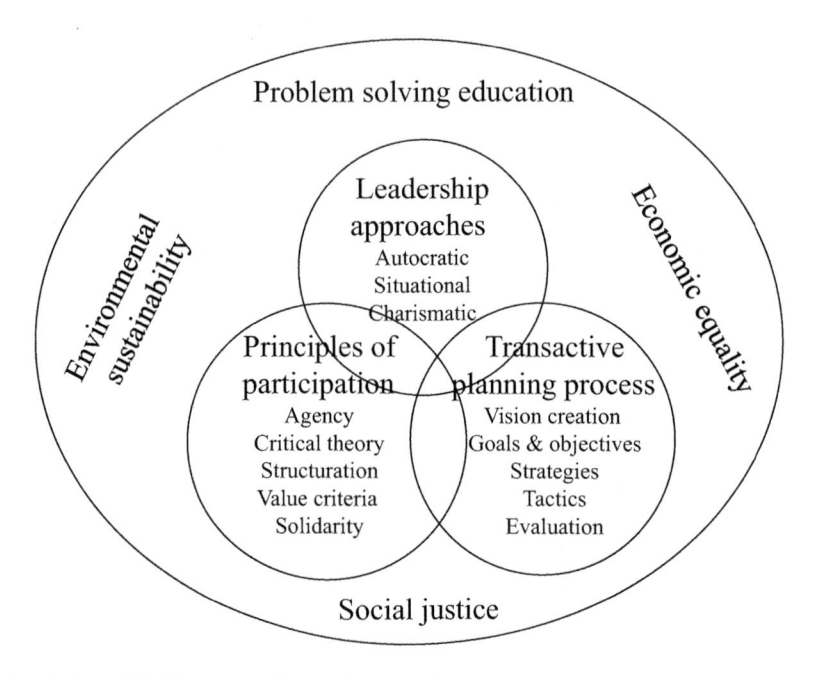

FIGURE 10.1 World-system change framework

the basic values on which the vision for the project can be based. The three inner circles in the diagram contain the principles and processes that provide the framework for action. The remainder of the chapter is dedicated to explaining the contents of the diagram more fully. That said, there is certainly more than enough room for the elements in Figure 10.1 to be modified, added to, deleted, or changed completely, as the public discussion progresses, and the details of the new world-system emerge. Figure 10.1 simply provides an initial framework for planning and action.

Before planning begins, it is important to set the context for that activity. Here we turn to the social theory outlined in Chapter 9 to establish some of that context. Those who might lead the process would need to immerse themselves in those models so that they would have a sound basis on which to proceed. The theories in Chapter 9 provide the principles and frameworks for dialogue on which eventual action can be built. These models would make clear the psychological and social conditions for responding to the problems presented in Part 1 of this volume and the various stages in the transition process. It is also important for the issues raised in Part 1 to be a part of the basic narrative on which the project proceeds. These issues constitute the substance and rationale for action. To not move forward on this journey would consign humanity and the planet to the vagaries of the market and the special interests that now control it. In my view this will not lead to the vision and the goals anticipated by society but simply provide more of the status quo leading to catastrophe. Change at the margins of what we now have as a world-system model will not be enough to change the basic trajectory of the present world malaise and, therefore, doom humanity to continual environmental and social breakdown, perhaps leading to the eventual destruction of humanity and the planet.

Critical to transactive planning for world-system change is vision creation. The vision should be based on the values contained in the outer ring of Figure 10.1. Detailing the values outlined in that part of the diagram would be highly participatory. It cannot be stressed enough that the people of the world must examine what it is they want the world to become and then set that vision as the fundamental principle on which planning and action proceeds. The inclusive dialogue this process demands can be based on Habermas' (1983) notion of 'communicative action'.

The eventual plan of action needs to be based on problem solving, not purely on an ideological argument. A problem-solving educational effort would need to be pursued at the local, national, and international level. A strong and legitimate international body is required to implement and oversee such a program. Unfortunately, the present world bodies, such as the United Nations (UN), the International Monetary Fund (IMF), the World Bank (WB) or the World Health Organization (WHO), do not possess the necessary worldwide legitimacy to carry out the task. So, either one or all of these organizations would need to be reformed and made legitimate, or something newly organized solely for the purpose set out here could be created. Before or in tandem with configuring a unit at the international level, work toward creating the eventual vision would need to begin at the local grassroots level. Problem statements and vision creation must rely not only on the emotional perceptions but also on a sound understanding of the problems confronting humanity now and likely to continue, into the future. A strong analysis of the problems such as the one presented in Part 1 would need to be packaged and dissected by the participants in the process. Out of that analysis would eventually come the vision. The vision needs to be aspirational and something on which measurable and verifiable goals and objectives could be built. Once the world's population comes to understand the true nature of the crisis and its implications for life on the planet, alternatives for the development of a new world-system can then be realistically undertaken.

An international task force legitimized by world governments, whether they be existing institutions such as the United Nations or some other entity, need to be mandated for undertaking the task of creating, implementing, and monitoring a vision-creation process for changing the world-system at the international level. Taskforces for dialoguing and consulting with the international body and organizing internal consultation processes would also need to be established by national governments. Feedback among and between all levels would need to be constant so the dialogue cascades into one another. It is an up-and-down process, not one way. Once a vision is created and agreed to at all levels, strong leadership from both the governmental and civic sectors of society would lead an educational campaign to explain to the general public the virtues of the vision and the rationale for action. If the consultation feedback process is strong enough during the vision-creation stage, the public should be fully consulted and in agreement with the content of the vision at this point. That said, there is always going to be a need to retell the narrative of democracy to continue to remind society where it is headed and what constitutes the overall goal. All subsequent policies would be evaluated through the lens of the vision statement. Finally, each country would need to create a planning task force to generate recommendations for their respective governments outlining processes, actions, strategies, and tactics for implementing the vision for the new world-system. Governments throughout the world would need to produce coordinated policies and, eventually, laws and regulations to implement the vision. International regulations would be created based on accepted national policies.

At the present time, the United States has created two competing, but very different, visions for moving forward into the future. The first is Trump's 'grievance society', where all questions to be addressed are seen as negative and external to the country or the fault of enemies within. At best this vision seeks to rectify past mistakes but says nothing about future goals. And, the lack of vision produces a limited range of options available for resolving the non-articulated problem. The second is an aspirational vision set out in the 'Green New Deal' which is progressive, developmental, and futuristic, but without clearly identifying the problem to be solved. It is not without its weaknesses. Certainly, the GND lays out future goals and objectives for society. It also hints at a possible vision for society without clearly articulating it. Vision statements are not a wish list but futuristic aspirational statements that can lead to social and environmental homeostasis. People need to engage in critical discussions to educate themselves on the pressing issues that makes world-system change so crucial. Dialogue on most, if not all, of these issues is beginning now in an uncoordinated manner throughout most parts of the world so there is demonstrated interest in solving this crisis. It is not enough for the leaders of this undertaking to understand the basis on which need for change is argued; they must also listen to the concerns of the public and their ideas on the subject if the project is to proceed in a coordinated fashion and with the majority support of the public. The problem to which the vision seeks to resolve needs to be clearly identified and understood by all. The aspirations and ideas of those leading the project need to agree with those of the population. If not, more dialogue is obviously needed until some form of consensus is reached. I say some form of consensus because unanimity is highly unlikely but there needs to be demonstrated sound support for the vision before proceeding.

Planning discounts the idea that structuralism and institutions, such as classical economics, predetermine the actions of people and how they transact everyday life. Institutional determinism must not be allowed to dictate the progression of civilization. That is not to suggest that the social infrastructure can be neglected or entirely discounted – it surely cannot. Institutions are the mechanism that facilitates social organization and action and must be treated as such. But it also needs to be remembered that social institutions are constructed and maintained by individuals in society. They are not entities apart from society. Transactional planning embraces the concept of negotiation based on the foundation of comprehensive research that provides the underpinning for the eventual plan. Transactional social planning recognizes the relationship between agency and society's institutions as it is described in the theory of structuration (Giddens, 1987). The basic principles of world-system change should

> begin with an ethic of indignation about the denial of human dignity and opportunities; incorporate human inquiry and understanding based on dialogue; [have] a focus on compassion and caring; and, a quest for social justice. These four cornerstones foster empowerment and social change to promote equity.
>
> *(Saleebey, 1990, pp. 29–40)*

As Wharf and McKenzie stress:

> Social planning is based on deep feelings of indignation about the denial of human dignity and opportunities. It incorporates humane inquiry and understanding into the process through dialogue and fundamental research. It is founded on compassion and caring for the human condition and environment, and, perhaps most of all it searches for social justice and is concerned for the advancement of humanity worldwide. Unlike most planning exercises that are focused either at the micro or mezzo level, the context for

world-system evolution is at the macro scale and requires a comprehensive transactive approach to the activity of planning. And, solutions should promote social justice.

(Wharf & McKenzie, 1998, p. 24)

What might constitute the value criteria on which a vision for a new world-system could be created? Planning for a new world-system can be founded on two basic principles. First is the complete understanding by the world's population of the impending global warming catastrophe and its consequences and, second, the need to rebalance the life- and system-world. Although it might look simple in written words, the task is daunting. Despite the complexity of this task, the human community has the capacity to undertake such a mission. Although identification of the problems to be addressed must form the basis on which the vision is created, people from across the globe need to be convinced that the new mission of society doesn't just solve the identified social and environmental problems but enhances their lives positively in discernible ways. Once the vision is set, then the discussion concerning the best methods and policies for achieving that mission can be debated.

It is critically important for the public to be engaged in the discussion at the earliest stages of the planning process. This can occur by establishing working groups in each locale. The results of these discussions can be fed up the chain. One of the most critical issues demanding attention is the problem of false consciousness. Much of what we humans accept as truth is handed down from previous generations. And while many of those truths were helpful for coping in earlier contexts, they may or may not be useful any longer and, in fact, may be detrimental to today's situation. These once useful truths are now considered false consciousness, and if our society continues to believe in them without verifying their present worth, they will detract us from pursuing the appropriate strategies for achieving our stated goals. The new system must be determined to be in the best interests of the planet and for all people who inhabit it and not for a small but powerful group of people.

As the process is stated here, I am under no illusions about how complicated and daunting accomplishing this process will be. In fact, there is a high probability that it might not be completed in time or at all and fail, although I shudder to think of the consequences. The world seems to be dividing, not unifying, as it slides in the direction of nationalism and isolationism. A unified approach to problem solving is compulsory if a new world-system is ever to materialize. The stakes are enormous if we continue to muddle along on the present course. We can see where that is taking us now. I hate to think of what will happen if action along the lines of what is presented here is not undertaken in some form. There are many notable and reputable scientists such as Stephen Hawking who think the planet has gone past the point of no return already and is doomed. I prefer to think that humans can adapt to the new environment and advance the human social system and at least partially repair the planet and keep it habitable.

Frequently overlooked is the necessity for constructing a management structure for carrying out the plan and achieving the desired results. What is most crucial is that the majority of people, including the governmental leaders and the powerful in society, buy into the necessity of world-system change. Certainly, many of them are very much concerned with the present chaotic system. They understand the dangers of the status quo. In addition to gaining buy-in by all sectors of society is avoiding the lack of recognition of what is often the most vulnerable part of the plan: implementation.

Democracy depends on the political arena where people vote for representatives who form the government every four years or so. World-system change will require greater involvement than simply voting every four years. The new world-system will require citizens to be much more active in governing the planet. The process of system change will include designing the economic operating system. This cannot be left to a single individual or small group of persons but must be approved by the general population. The population will need to be active in setting the values on which the new system will operate and in monitoring its continued progress. For large-scale social change of the magnitude envisioned here, building communication systems and developing social networks is critical. Technology provides the path for constructing such networks. Social networking is well under way through such avenues as Facebook, Twitter, and other social media platforms, although it is not always used well. The Arab Spring, BLM, Occupy, and the pro-democracy movement in Hong Kong are examples of the power of technological communication for building movements and coalitions of movements. Perhaps the next outstanding issue to be addressed is building the capacity among the world's population to engage in the process.

Social transition will require building capacity in society and in establishing networks of individuals and groups to lead the process. This will be a crucial step if success is to be secured. No doubt leadership will play a huge role in the next successful social transformation. Leadership has been an instrumental factor in societal change throughout history. Traditionally, it has been the rulers, priests, aristocrats, revolutionaries, or in some cases politicians who have initiated and facilitated change. Strong leaders from France, Great Britain, and Austria in the persons of Castlereagh, Metternich, and Talleyrand negotiated the Concert of Europe Agreement that ushered in the civilizing social European organization agreement after the Napoleonic Wars. This agreement had as much to do with developing the rules of conduct among these countries as it did with setting boundaries, and similarly with Churchill, Roosevelt, and Stalin after World War II. Understanding the role of leadership in the social transformation process will require continued exploration by politicians, academics, and the public at large.

Leadership must not act alone or in isolation to the larger public. Leaders must take direction from, and have the full confidence of, the populace as the process unfolds. The leadership must be held accountable and in check by the public. History is filled with examples of leaders who have separated themselves from their populations with disastrous results. I think of Hitler, Mussolini, and Stalin as strong leaders who led their population to disaster in addition to other extremely negative outcomes. Others such as Gandhi, Mandela, and Martin Luther King Jr. are examples of strong leaders who led their followers to great heights by being attentive to their needs in addition to other personal qualities. These successful leaders garnered legitimacy by being in constant communication with their supporters and by sharing in their values and desires. Part of building capacity in society is identifying and legitimizing a clearly defined group of individuals to lead the task.

The issue of legitimacy of the change agent and trust in the leadership of the movement is critical to social change. At times the leadership will need to be given license to make decisions on behalf of the total group. That said, this is often where the functioning of the movement breaks down and leaders must be very sensitive to boundary lines. Decisions requiring considerable dialogue and debate versus those that are sufficiently mundane that can be left to the leadership is clearly a contentious point, and leadership must read the situation carefully. Policy determination is the prerogative of all of us, but this does not mean giving any marginal group in the process a veto. Program design and implementation can be designated to a subgroup within the larger society. Evaluation of results will rest at the widest level of society.

I am certain that there will need to be at least two types of leaders in the world-system change project. In addition to those who are well known and qualified to lead a change movement because of their charisma, skill, and history of activism, there is also the need for sectoral leaders who can champion the cause. These champions will be some of the powerful and wealthy who understand the pending catastrophe and want to do something about it. People like Bill Gates, Tom Steyer, Warren Buffett, George Soros and others, who maintain some standing and admiration in the world for their philanthropy, will need to take a champion's role, particularly for getting their contemporaries and the public to buy into the criticalness of the situation and the necessity of world-system change. Many of these people are presently focusing on human issues not associated with their businesses but humanity's worst health and social problems so they are already showing sensitivity to society's need for change. They will be needed to spearhead an educational campaign that will encourage the world's population to understand the complexity of the situation and then to act on that conceptualization.

Giddens raises the issue of a 'motivational crisis' in addition to the usual issues concerning legitimacy of the leadership of the movement. The established leaders of society today continue to focus on yesterday's issues. Habermas and Giddens contend, and I think rightfully so, that societal 'conflicts no longer primarily concern the distribution of material goods, but rather cultural reproduction and socialization, and do not follow the established bargaining mechanisms associated with the unions or political parties' (Giddens, 1987, p. 307). This provides an explanation why unions no longer enjoy the support of the majority of workers they once did, nor are they seen as a force any longer for championing social change. The issues plaguing society are beyond economic and include environmental issues, migration, marginalization, and social anomie generally. There does not appear to be an institution such as the union where those seeking change can find a home that provides ongoing motivation and support. Certainly, such US groups as the Tea Party on the right and the Occupy movement and BLM on the left provide a fleeting and temporary home, but not a permanently structured institution that can grow in tandem with the increasing sentiment for change. That said, the Tea Party has taken strong root in the Republican Party in the United States and is said to control their political agenda, demonstrating just how influential a grassroots organization can become. Its success may have been somewhat dependent on considerable funding from the oligarchs on the right side of the political spectrum who share much of their worldview.

Capacity building and social networking are important parts of community development generally. The principles and processes embedded in capacity building and social networking can form the beacon for the birth and growth of new institutions that can fill the void left by the unions and the traditional political parties. The function of community development is to both educate the public and to accomplish a goal that is set and agreed to by the community. Perhaps the first strategy in this project will be the formation of social structures that can help create the institutions that will promote and accomplish world-system change. Continued development of the Occupy movement is one possible avenue as is further development of the BLM movement.

Community development is incorporated within the notion of alternative development. Friedmann suggests that alternative development ' . . . involves a process of social and political empowerment whose long-term objective is to rebalance the structure of power in society by making state action more accountable, strengthening the powers of civil society in the management of its own affairs. . .' (Friedmann, 1995, p. 31). Community development has also been defined as

> a process for empowerment and transformation. The focus on community development is to identify and resolve problems of a social, physical, or political nature that exists in a community in such a way that these conditions are changed or improved from the perspective of the community members.
> *(Reid & Van Dreunen, 1996, p. 49)*

The goals of community development include self-help, community capacity building, and social integration (Rothman, 1979). The most basic elements of community development are enabling the community to identify the problems inhibiting development; provoking a high level of participation and commitment by community members to resolve problems; stimulating the commitment and confidence among community members for engaging self-help practices and self-reliance; and maintaining community control of the process and outcome. To achieve the essential ideas contained in the notions of capacity building and social capital, leaders must be active in the community. In other words, the community must possess sufficient skill and social networks to undertake and complete the project. It is one thing to aspire for development but another to have the social capital and capacity to achieve it. The practicality of community development rests on the capacity of the community to act strategically and the level of social capital available for use by the general population. To achieve the level of cognition required by the population to engage in successful community development, people with special skill in these processes may be needed to help increase the level of understanding and capacity of the general population. Many of the civil groups that exist today promote community animation through experiential learning as part of their overall mandate and operations. Although these ideas are spoken about in terms of community, they apply equally well to larger scale projects at the national or international level, although they may require some alteration and an advanced level of sophisticated implementation. Community must be thought of as a community of interest not just in the geographical sense of the word. The basic idea is to cascade the work at the community level to the national and international setting and to coordinate actions across the globe.

The problems and issues that are recognizably present in the world today have spawned the development of social movements. Social movements come into existence when it becomes evident that the prevailing institutions of society no longer satisfy the needs or aspirations of a large part of the general population. Social movements identify the issues and problems with which society is confronted and develop plans and strategies for the amelioration of those concerns. It is possible that progress in addressing the social dissatisfaction that exists today can be resolved through negotiation with those in power or through non-violent revolution. Whether the problems confronting the world can be resolved through reform of the present system or through non-violent revolution is yet to be determined, however. Regardless of the intensity of the actions that are needed to develop a new world-system, planning is necessary for providing a logical framework for developing a rational course of action. After all, Homo sapiens means 'wise man' and we have the capacity to project and plan the future.

Fundamentally, the most effective method for driving social change from either the local or international level is from the bottom up and not top down. Bottom-up planning often happens when people become motivated because of some moral imperative or a basic feeling that things in society are not going the way they think they should. The following passage from a book written by Karen McCann titled *Women of the American Resistance: You Are the One We Have Been Waiting For* is an excellent example of a grassroots development that is making a difference:

> Like many, I was shocked and depressed after the 2016 election that I could barely breath; worse, I felt utterly powerless. Then came January 21st and the Women's March on Washington. I had just arrived back in Seville, Spain, where I live for a portion of every year; it's the home base from which I do my travel writing. Our little expat community had no one in place to organize a protest, so I sat at my computer and watched the images roll in from the world. For me, it was a stunning and profoundly thrilling to realize that there were *five million* of us – enough to mount a serious Resistance movement. We don't have to hunker down for the duration; we have the capacity to stand up and take a shot at changing history.

Ten days later my friend Sage McCotter sent me an email asking if I'd like to help her start a Resistance group in Seville. 'I'm in', I told her. Hours after our first meeting, I went to a dinner at an American friend's apartment, where talk quickly turned to politics, and the mood at the table alternately boiled with rage and plummeted into gloom. And then I mentioned that I'd just joined the Resistance. It felt as if an electric jolt shot around the table. There was a moment of absolute silence, and then everyone was talking at once, asking questions, few of which I could answer at that point. In the days that followed, every American in the room signed up to join our group, American Resistance Sevilla.

(McCann, 2018, pp. 4–5)

This small but highly significant group in Seville regularly plans demonstrations, marches, and protests against what they consider to be the antisocial policies and the politics of Donald Trump. They meet regularly to plan actions but also to engage socially and have fun together. They educate (experiential learning) themselves to the issues and the state of politics in the United States. Additionally, they hold periodic seminars and round table discussions on matters dealing with global warming and social justice. Although they are visible by their street marches in Seville (e.g., Trump's inauguration anniversary), they also mount less visible, but effective, campaigns. Each member has an app on their cell phone that directly ties them into their congressperson and senator in the United States, and they regularly send text messages and make phone calls to let them know exactly what is on their mind. There is also an educational component via a newsletter frequently sent out to their subscribers over the internet. I can't help but think there are many more of these groups within the United States and around the world that are well organized and acted strongly in the 2018 midterm elections that elected a Democratic majority to the House of Representatives. This group is also connected to the larger women's march organizers in the United States, and, therefore, see themselves as part of a larger movement as well (social networking).

The success of planning and implementing a new world-system will depend heavily on local, national, and international cooperation and coordination. I am under no illusion about the enormity of the proposition of world-system change. World leaders have not always been successful in gaining unanimity and cooperation or in synchronizing their goals and policies. I hasten to add, however, that there have been examples of cooperation and collective action. Most recent has been the ability of the Republican and Democratic parties in the United States to come together and create a financial rescue package for those affected by the COVID-19 pandemic. Apart from countries in the Middle East, there are no two bodies more acrimonious than the Republicans and the Democrats in the United States. Often, barriers break down when foes are faced with a crisis situation they must resolve. After World War II, much of the world came together to create the Bretton Woods Agreement for world reconstruction. The purpose of Bretton

Woods was to establish a system of rules and regulations to manage the world economy ensuring stability so it wouldn't slide back into depression at the end of World War II. Cooperation on climate policy attained at Paris in 2015 and subsequent agreements were established and approved even though these commitments are woefully inadequate, and their targets routinely not met.

Motivating the world to act is likely to be the largest hurdle to overcome in bringing this project to realization. Humans are naturally resistant to change because the present always seems more secure than an unknown but possibly better future. Human evolution has conditioned the human species to resist change and adopt behaviour patterns that have proven themselves in the past. It is like driving forward looking into the rear-view mirror. The change we need now, however, is change that is focused on redressing the environmental and social challenges that confront us, not on what has worked over the past. What is also different today is the speed with which change is required. Additionally, there are the powerful and wealthy who may have to give up their premium position in the world to save it. A lot will depend on the ability of humans to merge their emotional intelligence with rationality. All the planning techniques available will not be useful until the sentiment of the world's population fully understands the dangers of global warming to the planet and to all living things on it. Not until all peoples and countries of the world understand the severity and consequences of global warming can planning for its mitigation begin. And the timeframe for bringing this understanding about is short. The recent US government study on the subject projected a 12-year timeline before the planet steps over the point of no return and then it becomes a project of minimizing the consequences of global warming but not reversing the course. So, the first task is to substantially increase the intensity of the worldwide discussion on the subject. Not to be forgotten in all this is the pending crisis of social injustice that is fuelling much of the social anomie that exists throughout the world today.

Capitalism in its present form is not suitable for supporting the effort needed to address the ferociousness of climate change and global warming. It certainly has had a terrible record when it comes to dealing with social justice issues. It can be argued it has been particularly negligent in addressing poverty and inequality; in fact, it has been responsible for it. Capitalism is in the process of decline, and many are predicting it will implode. Any attempt to maintain the present world capitalist system will subject humans and the rest of the natural world to a losing struggle for survival. To use Hans Selye's GAS model as an analogy for the situation we face, the planet is being severely stressed by social inequality and environmental stressors and soon to be immersed in the adaptive phase of the GAS in an attempt to reach homeostasis. The adaptive phase will require major changes to our economic system if we are to fully address the pressing challenges. Whether or not we reach the exhaustion stage or achieve homeostasis will depend on our tenacity in meeting the world-system change challenge.

In this chapter I have attempted to outline a framework for moving society forward to a new social order. Previously in Part 1, I summarized the issues demanding change to the world-system. Other parts of this book hinted strongly that any attempt to make such a huge change in society must be undertaken from the bottom up. The plan and implementation strategy presented in this chapter feature a cascade-up model. With that in mind, communities need to undertake a social learning process that uncovers the details of the problems that threaten the planet and humankind. Discussion facilitators can be determined beforehand, and any special expertise, including issue specific reports, can be made available. These materials can be arranged and provided to participants prior to meetings and general assemblies. Only after local groups and individuals become sufficiently familiar and comfortable with the issues under discussion can agreement be reached on a vision for the future and moved forward to the next step in the process, which is to present the results of the discussion to the next jurisdictional level up the ladder. If agreement cannot be reached, then it must be assumed that more discussion is required before sufficient comfort can be generated. The definition of what constitutes sufficient agreement must be determined by the project participants prior to entering into any deliberations that may be structured.

Once the problems are identified and sufficiently detailed for discussion, the dialogue on the shape of the new world-system can proceed. This discussion would follow the basic format outlined previously. Straw votes could be taken to get a sense of the comfort level for moving forward in creating a vision and plan based on the discussion. Vetoes by any special interest group are not part of this planning design. Straw, but not binding, votes would indicate the amount of support for any proposal and enable the process to move forward or the need for additional negotiation. Again, the effort would feature and rest on the values of agency, communication and a bottom-up process.

Although the implementation strategy outlined earlier seems relatively straightforward and simple, it is filled with many potential pitfalls. It will be tempting to short-circuit the discussions in order to move the process along and get to the final stage. It is imperative, however, that each stage in the process be given enough discussion and learning time so that everyone in the process is fully informed on the issues and the potential positive and negative outcomes. People must buy into the process if the project is to be successful. The Tea Party and Trump Nation give us convincing examples of just how powerful grassroots organizing can be.

There may be some who would want to highjack the process for their own ends. This must be resisted at all costs. There is no predetermined outcome to the process even though I have hinted at some of the values and substance I think should be incorporated into the vision. Buy-in to the end product by the general population is critical if the process and outcomes are to be successfully reached. It must be understood that humanity needs to engage a new way of living and thinking that may be uncomfortable because it will be constructed on values different from those of the present world-system and not just a slight variation on them. That said, I think that many of the values that need to be in place to solve the current crisis are present in the population but may not have been articulated and become a unifying theme yet. We mustn't be afraid to discuss these types of issues or proposals openly and with force. Just look at the impact Greta Thunberg has had on the climate change debate within the younger generation and throughout the general population.

Given the urgency of the climate situation and some of the other issues presented in Part 1 of this book, there may be a need to act more quickly on some of them rather than wait for full implementation at the end of a linear process. Some of these issues already have agreement as to their make-up and resolution, at least partial resolution. In the case of climate change, we know for example that reducing the emission of CO_2 into the atmosphere is absolutely necessary. It should be relatively easy to get agreement on that fact. It follows, therefore, that such programs as a carbon tax or an equivalent can be introduced before planning for world-system change is completed. This approach emulates the action research model whereby actions are implemented incrementally and before the research is totally completed. It is an incremental implementation process. Those areas that suggest obvious conclusions before the research and planning project is completed are candidates for early action. They must not be seen as the final goal but only one small part of it however.

I recognize the enormous difficulty with what I am suggesting. Implementing the program that I have laid out in this chapter is extremely difficult – some would say impossible. The processes outlined previously all make very good sense in small-scale planning at the local, regional, or national levels, but at the world level it takes on a whole new grandness. I understand that what I am suggesting here is a huge undertaking and fraught with numerous difficulties and potential failures. It depends immensely on creating goodwill among the nations of the world and their leaders, which is not something we have achieved in the past. The largest stumbling block we regularly witness throughout the present climate change discussions is getting all the major and minor countries in the world together, to first agree to a program and then institute the sweeping changes that are recommended. Under normal circumstances, it would be prudent to act incrementally and institute changes on a gradual basis that would allow societies to adjust to those before moving on to introducing additional upheavals in the culture. Time, however, is not on our side. I keep visioning the possible calamitous scenarios if no change is made to our broken world-system, but we continue to focus on single issues and not on the basic problem that is responsible for all these social and environmental problems.

In addition to the problem of climate change and global warming, the issues comprising social injustice are creating a divided society, and the chasm is getting larger by the day. Present leaders who say they are concerned with the problems produced by technology, migration, and poverty are pointing to populism and nationalism as the way to address these issues, but these ideologies will only acerbate the unrest and frustration that exists in society today. The remediation of these issues requires as much attention as the climate change problem if society is to advance to the next level of civilization.

Some might suggest that it would be easier if each of these issues were addressed in isolation and perhaps in sequence. While this suggestion appears meritorious on the surface, let me reiterate that they are intertwined with one another, and successfully addressing each one is contingent on solving each of the others in unison. The major issues addressed in Part 1 are all part of the same system, and to fix that system, we must address the issues together recognizing they affect each other. Poverty, for example, is directly tied to climate change. Many of the poor across the globe are not thinking of conserving the environment or how much CO_2 they are spewing into the atmosphere but only of avoiding malnutrition. Many of the poor in the less-developed countries cut down trees for agriculture or firewood for cooking and warmth, which contributes to global warming. People who live adjacent to the Amazon continue to cut it down for agriculture and for its timber even though it produces about 20% of the world's oxygen supply and absorbs a tremendous amount of CO_2. By extension the unequal distribution of resources where some 42 people own as much wealth as the bottom 50% of the world's population is not only immoral but also a direct contributor to poverty and, therefore, climate change. Migration is a direct result of poverty and climate change as well. So, while it may seem prudent to deal with each of these issues in isolation to the others, let it be understood that each of them has a direct influence on the others, and all of them collectively have a distinct bearing on what needs to be accomplished in the end.

Bibliography

Carroll, S. 2017. *The Big Picture: On the Origins of Life, Meaning and the Universe Itself [Kobo version]*. Kobo.com.
Freeland, C. 2012. *Plutocrats: The Rise of the New Global Super-Rich and the Fall of Everyone Else [Kobo version]*. Kobo.com.
Freire, P. 2007. *Pedagogy of the Oppressed*. New York: Continuum International Publishing Group.
Friedmann, J. 1987. *Planning in the Public Domain: From Knowledge to Action*. Princeton, NJ: Princeton University Press.
Friedmann, J. 1995. *Empowerment: The Politics of Alternative Development*. Cambridge, MA: Blackwell Publishers Ltd.
Geuss, R. 1987. *The Idea of Critical Theory: Habermas and the Frankfurt School*. London: Cambridge University Press.
Giddens, A. 1987. *Social Theory and Modern Sociology*. Cambridge, UK: Polity Press.
Habermas, J. 1983. *The Theory of Communicative Action*. Boston: Beacon Press.
Kuttner, R. 2018. *Can Democracy Survive Global Capitalism [Kobo version]*. Kobo.com.
McCann, K. 2018. *Women of the American Resistance: You Are the One We Have Been Waiting For*. Cleveland: Cafe Society Press.
Reid, D., and Van Dreunen, E. 1996. Leisure as a Social Transformation Mechanism in Community Development Practice. *Applied Recreation Research* 45–65.
Rothman, J. 1979. Three Models of Community Organization Practice, Their Mixing and Phasing. In *Strategies of Community Organization*, edited by F. Cox, J. L. Erlich, J. Rothman, and J. E. Tropman, 25–45. Itasca, IL: F. E. Peacock.
Saleebey, D. 1990. Philosophical Disputes in Social Work: Social Justice Denied. *Journal of Sociology and Social Welfare* 17, no. 2: 29–40.
Shaar, J. 1989. *Legitimacy in the Modern State (2nd ed.)*. New Brunswick, NJ: Transaction Publishers.
Wharf, B., and McKenzie, B. 1998. *Connecting Policy to Practice in the Human Services*. Toronto: Oxford University Press.

11
MEETING THE CHALLENGE
From chaos to sustainability

The marginalized in society are demonstrably dissatisfied with the status quo, and many of the rest of us are becoming increasingly psychologically uncomfortable with the present world-system because of the looming, catastrophic consequences of global warming and social injustice that exists in society today. Those who benefit most from the present system seem to be the only ones satisfied with today's conditions, and if they are not, they only want to deepen the course we are presently on, not change it. There are, of course, many of our fellow citizens who are still unaware of the seriousness of the condition facing society today. Many of us do not yet recognize where the present world-system is taking us, and it portends the imminent collapse of life on the planet if left unchecked. It may seem daunting to most readers to contemplate the gigantic changes that are being advanced here, but the consequences of not making them are far greater than changing our present way of life. But where does one start?

Humans have faced crises before but maybe not with the same dire consequences that attend the present situation. What is hopeful is that on several previous occasions, civilization has transitioned successfully from one world-system to another, demonstrating its adaptability. There is no doubt that the present world-system is larger and more complex than previous structures. However, the earlier transitions must have presented as daunting a challenge to those civilizations as the current one seems to us today. Wallerstein suggests that world-systems transcend through rhythms until they encounter or build up internal contradictions much like Kuhn argues when he speaks of scientific revolutions. As Wallerstein puts it, ' . . . the system encounters problems it can no longer resolve, and this causes what we may call systemic crisis' (Wallerstein, 2004, p. 111). He goes on to say,

> (t)rue crises are those difficulties that cannot be resolved within the framework of the system, but instead can be overcome only by going outside of and beyond the historical system of which the difficulties are a part . . . The modern

world-system in which we are living, which is that of a capitalist world-economy, is currently in precisely such a crisis, and has been for a while now.

(Wallerstein, 2004, p. 112)

The world-system change model suggests that humanity has two options for responding to the crisis since the existing system can no longer function adequately. The first option is to attempt to reform the present system by tinkering at the margins, but often this simply extends the present condition until the crisis worsens. In present parlance, it is often referred to as kicking the can down the road, but eventually we run out of road to kick the can down. I believe we are now at the end of that proverbial road. As Wallerstein tells us,

> (t)his instability can lead to considerable anxiety and therefore violence as people try to preserve acquired privileges and hierarchical rank in a very unstable situation. In general, this process can lead to social conflicts that take a quite unpleasant form.
>
> *(Wallerstein, 2004, p. 112)*

Anxiety and fear have generated increased racism, xenophobia, anti-Semitism, and the division of society into tribes. Perhaps most to be feared is creeping authoritarianism and fascism. The swing to right-wing populism and nationalism we see today is an attempt to fix the problems that are currently incapacitating the system but are unlikely to successfully address the crisis. Populist nationalism gazes back in time for solutions rather than creating a new structure that directly addresses the current issues. And we seem to be glancing back over our shoulder a lot these days.

The second alternative is to create a new world-system that addresses the issues of global warming and the broken resource distribution system. What society faces today are new problems in a new context and they do not lend themselves to old remedies. If the world-system fails to transition to a new format, I suspect the human experiment as we know it will perish over time. Humanity has always managed to evolve to the next historical world-system and defeat the crisis of the time, and we should proceed to do that again. We do know that other species of Homo died out over the past, so we must remember that it can happen to us too. The social and cultural systems that help us organize our lives today are in a deep crisis and there is growing acknowledgement of that condition. However, the actions taken to address those concerns to date have been too feeble, and current administrations seem to be taking society off on a tangent leading to isolationism and social disintegration.

It would seem that few people are willing to examine the issue from a fresh point of view and to think outside the box. The virtues of capitalism are ingrained in the social psyche through years of eulogizing the ideology. However, humanity has transitioned through many different forms of social organization and various economic strategies. Do we really think that our civilization has arrived at the end of history and created the perfect social organizational architecture and there is no more advancement in economic arrangements to be attained? Are we really that arrogant?

History shows that societies change, and whether they naturally progress, or in some cases regress, is a matter for continued debate. Certainly, the /Middle Ages, specifically the period of the Dark Ages in Europe, would likely be judged as a regression, whereas the Industrial Revolution would, to most observers, be considered a progressive advance. No doubt any change will produce both positive and negative outcomes, but the Industrial Revolution was a positive innovation for most societies. The social indicators of health have increased over the last century, and industrialism has contributed greatly to the rise in the standard of living achieved by most modern economies. It is arguable that progress in the West has stalled, however, and may now be in retreat.

Societies from across the world have been operating on variations of capitalism for over the last two centuries. Nonetheless, it has changed dramatically since first enunciated by Adam Smith in his classic work *An Inquiry into the Nature and Causes of the Wealth of Nations* (Smith, 1771). Today, we are at the mature stage of capitalism which has transformed itself into a form of predatory corporate capitalism where wealth is concentrated in the hands of a few oligarchs, leaving many at the bottom of the economic ladder relatively impoverished by comparison. Figure 11.1 provides a graphic depiction of the change in capitalism over time, from its original conception to the present day.

As capitalism matured, democracy morphed into something quite different from its original form. In the beginning, capitalism was conceived within a democratic political ideology. That system relies on a social contract and an independent judiciary where disputes and protections can be resolved or found. When disputes arise in society, citizens and civil organizations exercise their right to make themselves heard through numerous means including, direct dialogue, demonstrations, civil disobedience, and, in the case of democracy, at the ballot box. Much of this discourse is embodied in the notion of free speech and the right to congregate. Corporations are governed by rules that ensure a level playing field between labour and the owners of capital. And, while this relationship has suffered from severe inequality from time to time, it has returned to equilibrium through the creation of such entities as unions that represent the interests of workers. The fact that union organizing has been on the decline over the last few decades is another indicator of the decline of democracy and the creep of corporatism. Supervising all this on behalf of the public, the fourth estate plays a watchdog role ensuring that the system does not get too out of balance in favour of any one group. When the government, corporations, or unions deviated too far from acceptable practice, the press is there to ensure that the egregious behaviour is brought to the attention of the public.

But something has changed from that idyllic view. We now see a corporate capitalism that is highly predatory, resulting in a few individuals worldwide owning as much wealth as the bottom half of the world's population. As a direct result of the concentration of wealth, many western governments are leaving the democratic governance structure and becoming plutocracies, where the wealthy own the politicians and dictate the policies and functions of government. In the United States we witness the politicization of the judicial system. The media is now in the hands of the wealthy and basically views the world through the lens of the owners

Meeting the challenge **195**

FROM

Capitalism
- Democratic governments
- Independent judiciary
- Access to political system
- International agreements
- Civic institutions (free speech/congregate)
- Corporate governance
- Free press

SYSTEM-WORLD

TO

Predatory Corporate Capitalism
- Creeping corporatism
- Financial corporate complex (responsible for 2008 crash)
- Plutocratic/authoritarian governments (Hungary, Poland USA)
- Politicized judicial system
- Media concentration (owned by oligarchs)

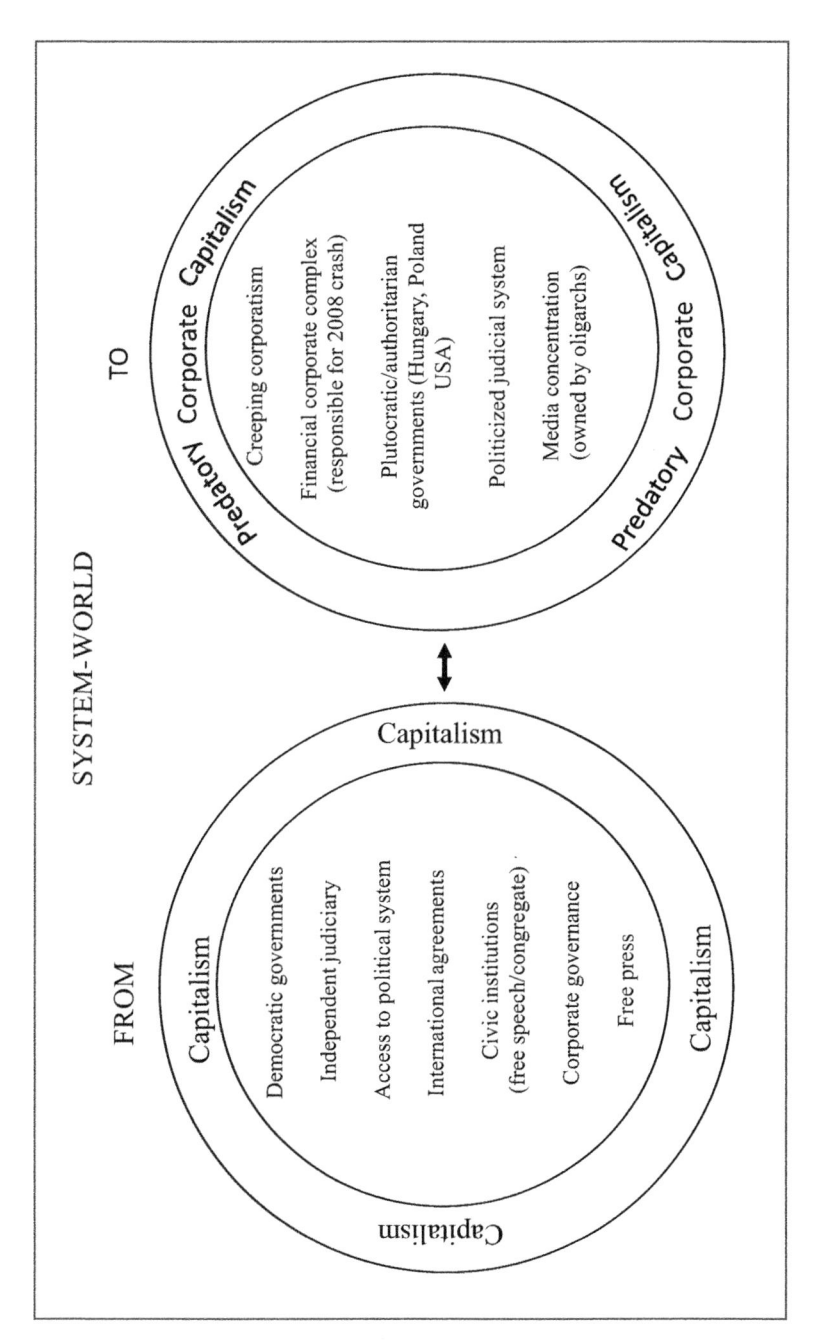

FIGURE 11.1 Capitalism to predatory corporate capitalism

of capital. Like any dynamic system, capitalism has changed over time. At present, it has been captured by a few wealthy oligarchs who have placed the primacy of profits ahead of every other social good, including a sustainable planet and just social system. As a result, some enlightened western governments recognize the deterioration in the system and are proposing measures to address the problems, but they are attempting to do it in the predatory corporate capitalist framework which is no longer equipped to adopt policies that will adequately address the issues.

There are many social scientists (Wallerstein, 2004; Streeck, 2016; McMurtry, 1999; Wallerstein, Collins, Mann, Derluguian, & Calhoun, 2013; Giroux, 2018; Reid, 2017; Hedges, 2018) pointing to the decline of capitalism, and some are musing about what might replace it. While many of them are adamant in their views about the end of capitalism, they are less certain about what will come after. Most analysts don't see Marxism being the replacement of choice; it too suffers from many of the same flaws as the capitalist system, particularly in relation to global warming and the challenge to human labour by AI. In fact, there is no accepted alternative waiting to be ordained, so a new world-system will need to be constructed incrementally through trial and error. Although socialism may go some distance in eliminating the economic inequality that exists in the world today, it says nothing about the issue of climate change and global warming. Socialism does not treat the externalities of production any differently than capitalism, leaving the environment vulnerable to the externalization of waste from the production of material goods and services, as in the case of capitalism. A new form of social democracy seems to be emerging and stimulating some discussion. It combines the sentiments of socialism (social justice) with the principles of democracy. It offers policies that may go some distance in addressing the issues with which this book is concerned, but an entirely new framework that directly addresses the matters outlined in Part 1 of this book will be needed to fashion and implement an entire new world-system.

If there is agreement about what follows capitalism's demise, the consensus believes it will be an extensive interregnum period. That is, a period of muddling through with many attempts at the development of a new system with some pieces of those experiments being successful and some not. Social learning in the form of trial and error with its dependence on feedback and an open two-way dialogue will play a vital role in this process. The process may produce many slightly different configurations for consideration. It could very well be that several systems will emerge at the regional level and then an overarching coordinating system constructed out of those fragments at the world scale. Some commentators and pundits refer to this action as deglobalization, but I don't think we can ever deglobalize completely in the new technological world. As much as some of us may not want to think, our trading relationships and supply chains are too widespread and complicated to return to isolated states. If the COVID-19 pandemic has taught us anything, it is that human-made borders are artificial, and many modern-day issues do not recognize those borders. Countries may group themselves in what seems like natural regional conglomerations and then come together to form an overarching structure that provides rules of engagement at the world level. Each region will need to figure out the best organizational structure for their society and then integrate those goals into a global framework that is clearly focussed on a collective vision for advancing civilization. That said, there will need to be a fundamental social architecture that coordinates all human activities in addition to policies that address regional and local concerns. There is great need for integrating responses to issues of mutual concern, and that list is getting larger by the day. Most likely a federated system will emerge that leaves the day-to-day operation of society at the local (regional) level but provides an overarching governance system at the global scale that is focussed on balancing the life-world with the system-world, thereby ensuring the sustainability of the planet and the social justice system. It is important to note that we live in a truly global and integrated world, and we share the consequences of what each of us does individually. The global nature of the problems society faces today is precisely why there needs to be a concerted cooperative global effort to solve the present world-system crisis. Prior to the establishment of a world-system, individual groups of people or small political units could solve their own unique problems without worrying about what other societies located in other parts of the world were doing. In fact, they may not have known about each other's existence. With the advent of communication technology and sophisticated modes of transportation, this is no longer the case. We truly live in a global village (McLuhan, 1967). Climate change, and other environmental pollutants do not recognize, or adhere to, political boundaries.

Mutually beneficial policies that address the needs of the planet and its human inhabitants will form the overarching approach, at first the regional scale and then at the global level. This is essentially what has occurred in North America and Europe with NAFTA (USMCA) and the European Union. When considering negotiations at the world level, the proposed Green New Deal (GND) may be one of those promising ideas that could provide the beginning point for discussion. Some of the aspirations presented in the United States' GND are compatible with many of the long-held policies in Canada and Europe, and the adoption of those symbiotic measures could be helpful to others as well. Here I think of universal healthcare, adequate environmental protections, social justice, and human rights policies. The policies laid out in the GND are not compatible with the present version of capitalism, so changes to the world-system will be required. Hopefully, this period of trial and error will be peaceful until the present state of predatory corporate capitalism can quietly end and a new satisfactory world-system emerge. The best-case scenario is a seamless integration and a transformation over a suitable period rather than an abrupt change. This may be too much to hope for given the stake a small but powerful group have in the present system, but one must maintain hope.

The present US administration's attempt to return to isolationism and nationalism can be considered a first attempt at constructing a new social system to address the grievances of their citizens, even though it has been an utter failure. In fact, the Trumpian approach while speaking the rhetoric of 'the people' is really an attempt to shore up predatory corporate capitalism using populist rhetoric. What many Americans fail to understand is that a proper functioning social democratic government is the answer to their problems, not the cause of them.

The GND was not originally intended to be a replacement for capitalism but might just have that effect given that many of its provisions are not compatible with the status quo. For example, it guarantees a well-paying job to all citizens. If adopted, this article would undermine the capitalist need for a permanent unemployed pool of labour to keep wages down and profits growing in a never-ending linear trajectory. Instituting a regime more like the GND would require that the dollars allocated by a plutocratic government to the wealthy and the private sector through tax cuts and other provisions would need to be redirected back to the public sphere to fund many of the proposed GND programs. This may be a drag on increasing profits and not welcomed by the capitalists but certainly benefit the mass of society and the environment.

Life needs to become less of an independent experience and more emphasis placed on collective requirements given that the consequences of individual actions have great significance for the collective. The idea of individualism must be embedded and defined by the collective and not seen as a competing idea. The population of the world is now too large for the type of unfettered individualism that has been extolled by many Americans in the past and we saw demonstrated during the COVID-19 virus crisis. As Etzioni (1997) tells us, individualism is embedded in a strong concept of collectivism. These are not competing ideas. Too much is now at stake to operate the planet in a purely competitive manner and without regard for the consequences produced by the actions of some that directly affect the majority. The GND asks the US citizen to contemplate the vital role that collectivism plays in creating and maintaining individualism. In most other western countries, the GND would be viewed as social democracy, but the US public has been indoctrinated for many years to fear the idea of communitarianism and socialism. Let's be perfectly clear on that point. The GND is not Marxist socialism but social democracy. The false consciousness perpetrated by the oligarchs that individualism cannot coexist alongside collectivism is now laid bare for all to see. The foundation for an individual free life is to be found in the collective with its laws and social constructions through which individuals operate safely and successfully (Etzioni, 1997). Individualism in a nationalist, authoritarian state is diminished by a winner take all mentality. It is a system where the strong overpower and trample the weak. Individualism in this context is anarchy. There are usually more losers than winners in this type of social system as history has repeatedly chronicled. Individualism was not originally intended for the powerful to trample over the weak, but that is what it appears to be doing today.

While the GND is a realistic attempt at fashioning a partial replacement for capitalism, it has some flaws of its own. Most importantly, it does not anticipate the replacement of labour by robotics and AI in the production and administrative process. Perhaps the greatest weakness of the GND is its focus on job training and the promise of good-paying jobs for everyone. This type of program is a holdover from the industrial age but no longer appropriate for the post-industrial society. Job training is a promise of a good-paying job in the future to pacify an otherwise angry mob, but those jobs are not likely to materialize. We already see an angry mob in those who chant 'build the wall' or 'lock her up', so the time for pacification is over. The technological future will all but eliminate the need for mass labour. The GND will need to change its focus from guaranteeing work to everyone to providing an adequate income to every citizen and in constructing multiple venues where individuals can fulfil their individual needs and find meaning in life.

Work and the job have been the space where individuals have connected to the capitalist system. For mass society, the income gained from work is the capitalist system. Work has been defined by modern society as the domain where most people find belonging and personal self-esteem, to use Maslow's scheme. For a lucky few, work and the job have contributed to their self-actualization. In modern society, the job often defines the individual. Because of the loss of manufacturing and other sectors to the developing world and to the new technologies, belonging to a workgroup and, subsequently, one's self-esteem has been lost for many. The attachment to work and the job may need refocusing for many who will no longer be able to find self-actualization through market employment. In a changed world-system, satisfying social and psychological needs through meaningful activity that is created and directed by the individual and supported through such economic policies as a guaranteed basic income is the future. This represents a huge leap given that governments and society have touted the virtues of work in the past as the only avenue for gaining life meaning. To make this leap, society will be required to reorganize and bolster its public services that focus on such areas as social work and recreation. This rededication of public services to this end is part of the suggested change to the social architecture of society.

Many modern states are now contemplating or experimenting with a guaranteed basic income (Reid, 2017; Bregman, 2017) to answer the question of how I will put food on the table. Self-fulfilment can be found through the creation of meaningful activities that are not solely focussed on the paycheque but on being productive to oneself and the community in which one lives. Humans have traditionally relied on outside forces for individual motivation and control such as the firm, church, or feudal landlord. The new world-system will change the locus of that motivation from outside forces to internal direction. The new world-system will emancipate humanity from the tyranny of the labour force. Jobs and work are a capitalist notion and will not have the prominence in the new world-system they have had in the capitalist, industrialist era. Think of the positive impact this type of social organization could have on the environment. Given the major environmental crisis society faces today, discarding the work-spend-work cycle and reducing the consumer society to something a little more sustainable would seem most prudent.

The human community may now be more willing to look at alternatives to the income/work relationship that has governed social relations over the past. Given the impact of the COVID-19 pandemic on labour, I suspect there may be considerable change to the structure of work for as long as it lasts in the emerging economy. The advantage of working from home is one small example of possible lifestyle change. Working at home may be a psychological bridge to a world without paid labour. The fact is becoming clearer, however, that advanced societies will need fewer and fewer workers in the production and administrative process as the new technologies take over the world of production. The question of how the fruits of that production will be distributed to the population is still a major issue to be addressed. This is a question that very few in society are willing to resolve and forms one of the most significant barriers to developing a new social contract for the future. As robots and other forms of AI take over the function of work, governments will be forced to tax them and not personal income as their main source of revenue. There are other possible arrangements that could facilitate the refocus of human life away from the traditional income/work paradigm. The lack of forethought on this issue may be one of the primary reasons we experience social unrest today. A new method of distributing the fruits of production in society could very well address the 'left behind' syndrome that so many people in the developed world are experiencing.

Why is the struggle between Trump's version of authoritarian nationalism and the left's program of social democracy (or the United States' version of it in the GND) important to the rest of the world? Over recent history, the United States has presented itself as the exceptional version of democracy and capitalism. Many other parts of the free world are also in a social struggle and in turmoil, much like in the United States, perhaps not on an identical path but surely on the same trajectory. All movements across the world suffer from the same paralysis. For lack of a better description, it has been termed 'grievance politics'. Those left behind in the globalization project are looking for someone to blame and that anger has been directed at liberal democracy itself. The right-wing populist movements across the globe have tuned into that anger but have not projected any positive policies to move their countries forward in a constructive manner. The program of the populist/nationalist movement is constructed on emotional intelligence, not rational thought. The prescription is simply to shut itself off from the rest of the world. It has been constructed on negative grievance-based politics such as those found in Trumpism and Brexit that desire a return to some bygone era rather than facing the reality of the future. A forward-looking program such as the GND is offering something different in a rationally based positive program that speaks to the reestablishment of the disenfranchised and marginalized to their rightful place in society. The proposals contained in the GND offer positive developmental programs in contrast to the ideological and emotional content of the right-wing populist message. The rest of the western world will take notice and convert to this path if it shows signs of success. Like it or not, the United States has often set the direction in the world, not only in establishing new cultural patterns but also in advancing the social architecture of society. The traditional leadership of the United States may be in decline, however. Their inability to advance rather than retreat may provide an opening for another society to show the way.

I should point out that while I have presented the GND as an alternative to the Trumpist politics of today, I do so only to suggest there are alternatives being offered that are sufficiently radical to move society toward a new world-system. By itself the GND is not sufficient to construct a new world-system. There will need to be many more component parts, but the GND is a good start from an aspirational perspective. It certainly provides an alternative to the Trumpist nationalism we witness today. I am quite sure there will be other equally compelling proposals to come along once the world understands the necessity for moving in what may seem at the moment like a radical direction.

World-system change is underway and has been since the 1970s. However, we seem to be in a chaotic phase at the moment. As Giroux suggests, '. . . the path we are on will lead to more misery and conflict' (Giroux, 2018, p. 31). If we are to form an egalitarian state that is mindful of the physical environment, and takes a more caring attitude toward others, we will need to implement a process that focuses on the needs of the entire community and not just on the desires of the few. The present path does not seem to be leading to a suitable destination but one that is antisocial and divisive, pitting one sector of society against another. There are many avenues and pitfalls to be avoided as we move forward, but forward we must gaze, not yearn for some nostalgic past. Giroux continues, '(o)ne place to begin is with reason and truth, and how fundamental they are to creating critically engaged citizens and communities' (Giroux, 2018, p. 25). Truth for many of our political and corporate leaders has been in short supply recently.

Donald Trump and Nigel Farage, among others, have given legitimacy and voice to the grievance society. They have capitalized on the disenchantment of the marginalized members of society and focused on exaggerating those fears rather than offering positive policies to address their issues. The grievance society, in and of itself, will not improve conditions for those who feel left behind. The urgent need for change, given the crisis produced by global warming and the increasing social unrest that is gathering momentum, is self-evident. Carbon pollution is not abating sufficiently to avoid catastrophe, and the inequitable globalization of the economy is now producing a severe backlash in societies across the globe. We may be witnessing deglobalization through the rising threats of nationalism and populism that are gaining strength in many democracies throughout the world when, in fact, we should be strengthening international commitments to act positively and collectively on these critical problems. Deglobalization is built on the suspicion and mistrust of others. During the COVID-19 pandemic, I heard leaders say that we are in this together. We need to extend that sentiment to a number of issues at the international level and not end it at country borders. This is particularly worrisome given the recent threats to the international finance system and manufacturing supply chains that have been productive and systematically structured over the last few decades. It is not completely understood what damage unwinding these systems will produce across the world, but it cannot be good. I make this comment confessing that I was skeptical about NAFTA when it was first introduced in 1994. (NAFTA was a replacement for the Canada–US free trade agreement that had been in force since 1988.) Since that initial skepticism, the international economies that have embarked on free trade agreements have consolidated their economies to such a state that to unravel them now would do irreparable harm. How Brexit plays itself out over time should be instructive on that matter. Deconstruction of the international economy will not take us forward but is a retreat to isolationism that will lead to an economic, environmental, and social downward spiral. The distribution of the benefits these systems produce needs to be rethought, not the systems themselves. Restructuring, not eliminating, these structures seems most prudent. The reform needs to be contemplated within a philosophy of social democracy and a new world-system not right-wing populist nationalism. The world acting collectively can enhance human rights policies and promote environmental and social justice to a level that will strengthen economic trade policy. The system needs to be based on the basic philosophy offered by Polanyi (1944) who argued that the economy should serve people rather than people serving the economy. This measure would rebalance the life- with the system-world. The erosion of the life-world is due in part to the takeover of the system-world by the monied class. The system-world has grown and come to overwhelm the life-world, damaging the social structure. Figure 11.2 outlines this situation graphically.

The higher order needs, beginning with belongingness and self-esteem, have been under attack for many citizens who, until now, were able to satisfy many of their psychological and social needs through work in the capitalist system. The GND is a direct refocus on bringing back the balance between the life- and system-world. The aim is to bring the life-world up to strength and to refocus the system-world, not diminish it. Figure 11.2 offers policies that are obvious for rebalancing the life- with the system-world. This is not an exhaustive list and there are likely many more appropriate possibilities.

Until a world-system paradigm shift takes place, membership in resistance groups such as the Tea Party, Trump Nation, or the Brexit movement will continue to provide an avenue for group belonging and self-esteem to reassert itself, at least temporarily, for some of the people who have been marginalized in the new economy and for those whose work no longer provides them with the higher order satisfactions outlined by Maslow. As the contents of the circle on the right in Figure 11.1 illustrates, the system-world has become dominated by the financial corporate complex, the conversion of democratic governments to plutocracies, or, worse, autocracies, and media concentration owned by oligarchs who provide an ideological interpretation to their reporting. And much of their reporting is designed to simply inflame their partisan audience to keep them in the grievance state of mind.

Figure 11.2 offers some policies that would help bring the system-world back into balance with the life-world. The left side of the graphic contains a larger broken circle suggesting that the life-world should expand to the size of the system-world. Some of the policies that could increase the health of the life-world are implementing a 70% marginal income tax for income over 1 million dollars a year (or some designated figure of choice), which would not only reduce the hegemony of the wealthy class but also provide more funds to deploy for life-world functions such

Meeting the challenge **203**

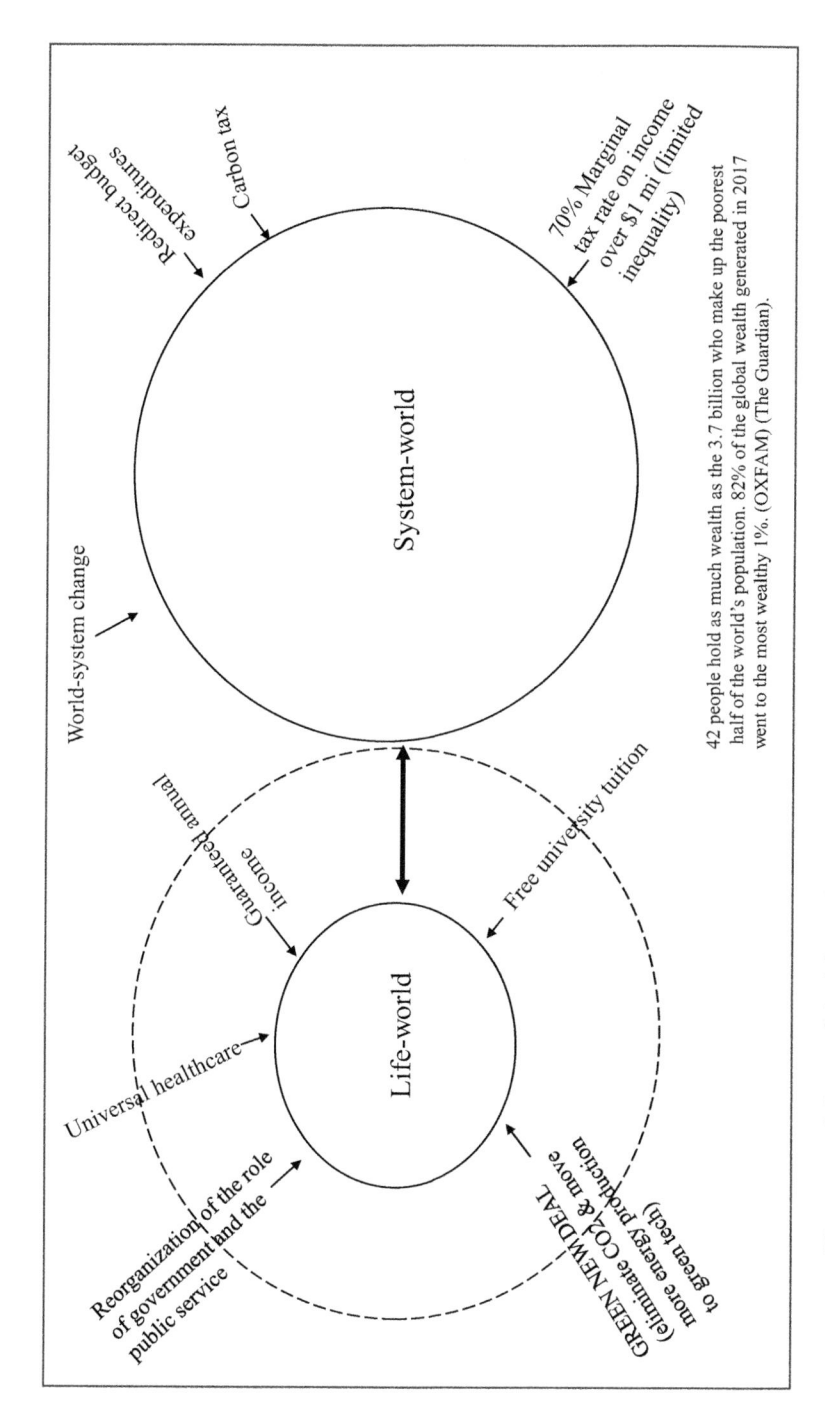

FIGURE 11.2 The present balance of the life- with the system-world

as healthcare. Reducing military budgets and redirecting those savings to programs proposed by the GND would help increase social solidity. Some of this redirection could also be used to contemplate a guaranteed basic income (GBI) for those who find themselves permanently out of work. A rededication of parts of the public service to assist the unemployed to make the social and psychological transition to a life outside the workforce would also be required. The revenues raised by a carbon tax could positively affect the environment by reducing the amount of CO_2 that is now being spewed into the air and provide revenue for use in creating new green technology for energy production. Let there be no mistake, the collective largely through governments will need to play a greater role in the new world-system than in the present version. The COVID-19 pandemic demonstrated the inadequacies of the private sector in a time of crisis. The capitalist system focused on what was good for business, not what was in the interest of the general public, and, as the pandemic showed, society was not prepared to meet the challenge because of its singular focus on economic growth. It has become clear that many governments let planning and preparation for fighting a pandemic slip because all attention in the modern era has been placed on growing the economy that only benefits the few. This unacceptable state of affairs will need to be rectified in a new world-system.

Under normal conditions, the system-world is constructed to facilitate the smooth operation of social relations. Figures 11.1 and 11.2 suggest it has become dysfunctional in that mission. A clear indication of the dominance of the system-world and its relation to the life-world is the recently published OXFAM report (Elliot, 2018) that claims that 42 people hold as much wealth as the 3.7 billion people who make up the poorest half of the world's population. In 2017, 82% of the global wealth generated went to the wealthiest 1% (Pimentel, Aymar, & Lawson, January 2018). The system-world has come to dominate the life-world and that is the locus for much of the social unrest and environmental problems that exist in society today.

The life-world is portrayed in Figure 11.2 to be much smaller than the system-world and needs to be expanded to the size of the broken line circle. This imbalance is now recognized by some and starting to be discussed in the US political system by many of the newly elected members of the House of Representatives. The discussion is beginning in other parts of the democratic world as well, mainly but not exclusively because of the shock of the COVID-19 pandemic. Some of these new politicians have argued for policies that could bolster the life-world and for additional actions that can reduce the dysfunction of the system-world (see Figure 11.2), but the world is still a long way from actualizing those ideals. Critical to this discussion is the involvement of the general population that suffer from the present imbalance. Politicians can't do it alone; it will take support from the population to make such change. But the COVID-19 pandemic may provide the catalyst for change. The health crisis produced by the virus may have provided the experience needed for the mass of society to contemplate a better world. We already hear the phrase 'new normal' in the post–COVID-19 lexicon. Just what that new normal turns out to be is anyone's guess, but surely it will be something different than what it is now.

Giddens summarizes Habermas when he states,

> [T]he colonization of the life-world has destroyed traditional bases of communicative action, without replacing them with forms of post-colonial rationality that are required to couple the life-world to the range of activities controlled by expanding economic and political steering mechanisms. . . . The new conflicts, and associated social movements, derive from problems that can only be resolved through a reconquest of the life-world by communicative reason, and by concomitant transmutations in the normative order of daily life.
>
> *(Giddens, 1987, pp. 306–307)*

In fact, there may be some argument that change to the social architecture often occurs when the balance between life-world and system-world becomes extremely distorted, particularly when that imbalance is in favour of the system-world as it is now. We are witness to that extreme imbalance in western society today. Will this extreme disparity be enough to motivate mass society to act? Let's hope so.

The system-world is generally comprised of the dominant structures in society such as big business and national governments. President Eisenhower dubbed it the Military Industrial Complex. More recently, pundits have referred to it as the Media Celebrity Complex. Some go as far to argue that Donald Trump has modelled his administration on the TV reality show that he hosted for many years before embarking on a political career. Modern humans have become habituated to that model in their daily lives, so why shouldn't reality TV carry over into real life? Life now imitates art, if you can call reality TV art!

Habermas critiqued how the media and the system-world arranged communication to be top down rather than an equal dialogue between all parties. Unequal weight has been given to institutions. Society's institutions often consider themselves indispensable to the functioning of society and to some extent they are. The problem arises when the prevailing attitude of the elites in control of these institutions suggests that mass society is there to serve them rather than the other way around. Oligarchs spend considerable amounts of money to convince the masses that the present social organizational structure, perhaps an even more exaggerated version, is in everyone's best interest, which has been proven to be clearly not the case. Habermas (1989) argues, and rightfully so, that the foundation of critical theory depends on a system of communication that makes all parties in the dialogue equal. As Swingewood notes when reflecting on Habermas' ideas:

> Individuals possess the capacity for self-reflection, understanding and knowledge. A theory of emancipation is built on a model of communication not production, with Habermas positing as an ideal structure of undistorted communication the 'conversation of free citizens', action oriented towards truth. The life-world is the sphere of free and equal discourse, of rational understanding and a 'normative consensus' that flows 'from the cooperative interpretation processes of participants themselves'.
>
> *(Swingewood, 1991, p. 294)*

Critical theories not only embrace Giddens' notions of agency but put great emphasis on actors cooperatively participating in their own emancipation. The life-world/system-world structure can provide a critical lens for evaluating all proposals for change to the world-system.

The world-system change project must denounce politics as usual and focus on ideological transformation, not simply the replacement of one mainstream political party with another. Although this writing has focused on the inadequacies of the Republicans under the control of Donald Trump, the Democrats represent the same monied class as their political foes. Undeniably, the Democrats want to improve healthcare and other social programs for the average person, but they do operate within the same political framework as the Republicans. Money has a strong hold on the Democrats as well as with the Republicans. This condition is more or less true in other democracies around the world. Some argue the Democrats have become the party of the elite, not of the people. Remember, it was a Democratic president who terminated the Glass-Stiegel Act that since the Great Depression of the 1930s ensured fiscal responsibility on the part of the finance and banking system and would have limited the carnage of the 2008 financial meltdown if it had still been the law of the land. And it was the same Democratic president who reduced some of the social welfare programs on which poorer citizens depended for their wellbeing. The Democrats are beholden to the wealthy for their political success as well as the Republicans. Perhaps to a lesser extent than in the United States, money seems to have become too dominant in the politics of the democratic world. World-system change will not occur as long as there is politics as usual. The rise of Donald Trump and Brexit is a reaction to politics as usual, even though a misguided response. The present political system operates on meanness, grievance, and deceit, and all mainstream political parties are beholden to the wealthy class for their eventual success. No new system can come into effect until society addresses these considerable flaws. On a more positive note, some members of the Democratic Party in the United States are attempting to become independent from the moneyed class by not accepting their largess. Unfortunately, many of these politicians are being termed 'the far left' or 'the socialists' to discredit their motivations. History is likely to judge them as true patriots given their unrelenting focus on the many problems presented in Part 1 of this book and their rejection of politics as usual.

To avoid a downward drift in society to an uncertain future simply through kneejerk politics that promises a past life that cannot be regained, society must move forward in a planned way. This plan must address the major problems of climate change and the economic marginalization and social alienation of many people across the world. As Giroux tells us, the present power structure ' . . . consistently prioritize corporate power and financial interests over social injustice and the common good. We must resist efforts that equate corporate commercialism with democracy' (Giroux, 2018, p. 152).

It must be remembered that humans depend on the biosphere for life. If exhaustion was to occur in the biological sphere, changes in the social domain will be of little consequence. That said, what happens in the biological realm is very dependent on how society arranges its social structure. There is symbiosis between the social and biological domains. No matter the present state of that symbiosis, the current crises will lead to biological and social exhaustion unless we can change the path society appears to be on at present. As we approach the end of the resistance phase of Selye's model, we will either adapt and change the circumstances inherent in the present world-system or lapse into the final stage of Selye's model and collapse into the exhaustion phase. Time is running out to mount an adequate response to the present crisis. Dramatic CO_2 reduction and the immediate reorganization of the elements of the predatory corporate capitalist system requires attention and transformation from the present toxic state.

There are active citizen organizations that are dedicated to reducing the stresses caused by the problems and issues summarized in Part 1 of this book. They are already active and growing in numbers. While these efforts are noble, what is often missed by single issue activists is understanding that these problems are interlinked and solving one without addressing the others will not change our overall welfare significantly. The conjuncture of global warming, replacing human labour in the production process through technology, poverty, and economic inequality, and the issue of migration and the clash of cultures are the complex root causes of the crisis in civilization today. Unintentionally but emphatically, the issues mentioned in the last sentence have contributed to the backlash against globalization and the apparent attempt to deglobalize. Globalization in and of itself is not an inherently evil organizing strategy; in fact, the world must become more rather than less integrated. Technology, particularly communications, is too dominant in our culture to wall it in and isolate ourselves from the rest of the world. Air pollution, for example, does not respect borders; it is a global phenomenon. Certainly, diseases such as the COVID-19 virus is a clear indication of the need for global integration in order to stave off catastrophe. It is the terms on which global integration is built and implemented that matters. The present focus of globalization has been too narrow and only concentrates on the financial and economic systems. To combat this breakdown in the globalized system, the crises presented in Part 1 of this book must be addressed simultaneously and in a coordinated manner, and solutions to these problems must be created considering the effect each has on the other. Action on one issue cannot be taken at the expense of the others. All these issues must be given equal priority. Until we act on all these problems concurrently and create a new acceptable world-system that can sustain the environment of the planet and provide social justice to all, the general crisis will not abate, and civilization as we know it will continue to hang in the balance.

Right-wing populism is not created by capitalist economic institutions *per se* but is a reaction to their failure to provide social justice to all people. There has always been an uneasy relationship between labour and capital, but labour was almost always able to deflect the worst intentions of capital and keep it on track. Capital over the recent past has neutralized labour and the other democratic institutions that kept capitalism in check. It was when capitalism became predatory and dominated by corporate interests that mass society became disenchanted. People are looking for explanations for their plight, and the rhetoric of right-wing populism and nationalism has captured their fears and anxieties.

Donald Trump did not generate the large numbers of people who comprise his movement. He taped into their anguish and fears and became their lightning rod. The fact that the promise of democracy has betrayed the masses has led us to Trumpism and Brexit and the other disenchanted leaders who populate the world. Voters have become alienated from their traditional worldview and now believe it is no longer operating in their best interest. Liberal democracy is losing its appeal, and there is heightened detachment among the population, and the search for a new form of politics and governance has begun. During this anxious time, variant forms of strange symptoms appear in the public discourse and system, and the search for a new world-system may produce dead ends before the correct course is designed. It can be argued that present society, particularly liberal democracies, are in a state of chaos and that is why we are now witnessing the rise of authoritarianism, isolationism, and populism (Giroux, 2018).

Authoritarian leaders ride the anxiety of those suffering from anomie in society who have legitimate grievances but lack constructive leadership to channel that dissatisfaction in a constructive manner. Authoritarian leaders operate best in a grievance society. While they identify with the grievances of the disenfranchised, they do little to rectify the problems and, in most cases, acerbate the tensions and anxiety by blaming their grievances on people or institutions that have no role in their manifestation. In fact, they thrive on it and would not have a platform if those grievances were addressed and resolved. For decades, the United States was able to project its internal ills onto the Soviet Union. Since the collapse of the 'evil empire', the immigrant and other countries in the world that some believe have taken advantage of the United States are now the favoured target. Authoritarian leaders focus their efforts on finding externals for society's problems rather than creating policies to alleviate the real internal deficiencies. Leaders must attack social and environmental problems by reaching out to the best minds and then carry out wide-ranging consultations with the public before determining what policies to implement. Appropriate leadership will reach decisions grounded in sound research that is fact-based, not purely on emotion or ideology.

Politicians and the power structure should not attempt social change in isolation to the public. People from all walks of life and social classes share the fate of the planet and, therefore, are entitled to participate in problem identification and rectification. Perhaps, most importantly, they need to be instrumental in creating the vision on which the plan will eventually take shape. At the present time, politicians and political systems are under challenge and do not possess the legitimacy to take bold decisions or actions on their own without engaging in widespread consultation with the public. Consultation in this context follows on the lines of what Habermas (1989) intends when he speaks about communicative action. The participation strategy must also focus on rebalancing the life- with the system-world. The needs of mass society have been forgotten while the focus has been placed on a narrow set of institutions and not with the just distribution of the prosperity produced by these organizations. Changing the world-system will require a planned approach and not rely on some invisible hand that supposedly guides the economy. There is no naturally functioning economy that operates on its own without human intervention. Humanity must plan its way forward.

There are four simple, but difficult to achieve, stages to system change. The first stage is to recognize that it must be planned; it won't just happen by itself in some mystical way. Associated with planning of the economy, the public will need to be educated to the fact that planning is not 'socialism' but collective, society wide, goal creation, and resource allocation within various sectors of society. Government involvement is absolutely necessary in the operation of society. Additionally, the oligarchs must no longer be allowed to control the system. Further, it needs to be acknowledged that there is no ready-made system out there waiting in the wings to replace a failed capitalist system. Second, the current social organization strategy – predatory corporate capitalism – must be seen by the masses to no longer serve the true interests of humanity or society. Third, mass society needs to come to understand that we are in a downward spiral and not just in a normal recoverable downturn. That recognition has been reached by the Tea Party and Brexiters but not necessarily by the general public, although that day seems to be fast approaching. Finally, society needs to identify and design alternative methods for organizing its affairs. At the end of this stage, society needs to legitimize the new paradigm and collectively implement a plan in order to move to the next elevated stage of civilization. The Trump experiment will be short-lived, and the GND looks, at least in the initial stage, as if it has promise for beginning the development of a new paradigm. At the very least, it represents a new course for society to ponder and maybe pursue. But analysis and planning are the key to moving forward.

The planning process must, in the first instance, be transactive and identify the value criteria on which the effort will be based. All too often it is assumed by those in charge of the process that everyone engaged in it shares their values and goals. This is a fatal mistake. Even though time is short for addressing the issue of world-system change before catastrophe befalls, a thorough discussion of the values on which the new system is to be constructed needs to be given priority in the early stages of the process. It is recognized that the size of this undertaking is massive and complex. There will surely be critics of any attempt to move away from the status quo. Simply making false and outrageous statements about any plan that is put forward in order to continue to engender fear in the populace will not alter the need for creating a new paradigm. Too much is at stake, and there appears to be some movement underway to move toward a new paradigm albeit somewhat antagonistically. Conferences on climate change, such as the Paris Climate Accord and the follow-up to it in Poland in late 2018, are evidence of this movement. As important as these initiatives are, there is still an urgent need for a bolder and larger effort, and greater participation by all countries in this process.

It is critically important for all people to educate themselves about the issues to be addressed and the potential solutions to the problems they face. The education process cannot be left to chance but needs to be well planned and implemented. It needs to be recognized that the vision to be achieved or the process for achieving it is not the sole prerogative of the leaders of the process. It is critically important for the public to be heard and involved in the creation and enunciation of the values on which the new world-system will be dedicated and not just with getting on board at the solution stage. Many processes, such as the one intended here, fail because the solutions to problems created by the powerful do not speak to the needs of the masses. An inclusive strategy must be designed before any vision or plan of action can be developed.

Processes will need to be created and put in place to facilitate this social transition. This will undoubtedly require social change through increased rational thought and collective decision making. Guided change will require that society collectively and comprehensively examine the problems that jeopardize the human condition and fashion a cohesive transactive planning process that solves the problem. That said, it is not out of the question that a revolution will take hold if change is left to drift rather than acted upon by rational forces. We see the makings of revolution in many western countries today by the rise of nationalism and isolationism, and by the emergence of potential leaders who appear to be strong on law and order, anti-immigrant and xenophobic, which are the symptoms of the problems but not the problems themselves. We are also beginning to see a reaction against the tactics used to oppress the marginalized by those in power.

It may be that we have embarked on a path to nowhere in an attempt to solve present social problems by turning inward and endeavouring to retreat to what some people regard as a previous golden age. Again, this is simply the decoupling of problems from potential solutions often leading to a mismatch between the problem and a set of potential solutions. Both the substance and the process that has led to the populism and nationalism we witness today has not been methodical or inclusive; it has been heavy handed and ill conceived. The present populist focus will simply lead to more social upheaval and disappointment by those who feel dragged down by the present system. What is required today is new and progressive thinking that rationally analyzes the present condition and constructs a new social contract that addresses the issues challenging present society. Short sound bites by our political leaders instil fear and anger in the population but do not provide any worthwhile information or analysis on which to act. The path forward must be based on scientific fact, inferences made on those facts, and a rational debate that generates appropriate solutions to the problems we face. The new social contract will need to resist engaging old worn-out remedies that probably didn't work even when first proposed and have less chance of working now. There may have been an acceptable larger margin of error for making mistakes in earlier times than there is at present. The state of the environmental and social world is so precarious and fast moving that a single error in judgement today could have disastrous results for all of society in the future. Such harmful decisions can be seen in Trump's refusal to sign the Paris Climate Accord and the dismantling of the environmental protections that were constructed over many years by the US government's Environmental Protection Agency.

Technology is changing at a fast pace and the social infrastructure required to organize life needs to pick up the pace of change in order to cope in a timely fashion. Technology is now 21st century and moving very quickly to become 22nd, but our social system is still immersed in the 20th and struggling to catch up. Society's technological advance is now outpacing the ability of our social institutions to change. Also, it is quite apparent that our political and social systems are not coping well with climate change or poverty reduction either.

What North Americans do will undoubtedly affect those living on the other side of the world, and vice versa. This points out the necessity of seeing the paradigm shift in global terms and not solely as a nation state issue. Most of present-day problems will only be resolved through multilateral cooperation where the problems faced by historical societies were often determined by an individual country or smaller group initiatives. What is critical to this discussion, however, is the destructiveness of the present geopolitical conflict. Given the deteriorating environmental conditions of the world and the inequality that exists in many, if not all, parts of the developed world, the present conflictual approach to international relations does not bode well for successful geopolitics. What is needed at this time in history is cooperation among nations, not conflict. The problems facing the world today cannot be resolved but only exacerbated by relying on a struggle for power between two of the world's largest economies. The only successful way forward is for cooperation between and among nations to solve these intractable issues and not for them to engage in out-and-out conflict for power and domination. In the end it will be how power is used and for what purposes, not its possession that will determine humanity's success or failure to advance to the next level of civilization. Although we appear to be moving in the opposite direction today, eventually we will need to construct a global framework to set the basic rules that govern the planet. The world-system, not the nation state, will become the focus for analysis in the future.

To move the process of world-system change forward, we must first deal with the creeping notion of system inevitability and eternity. Snyder defines the politics of inevitability as 'a sense that the future is just more of the present, that the laws of progress are known, that there are no alternatives, and therefore nothing to be done' (Snyder, 2018, p. 11). Democratic society is not there yet, but those trending toward populism and nationalism are taking a bold but ill-conceived step in that direction. The politics of inevitability ends with the politics of eternity, that is,

> one nation at the center of a cyclical story of victimhood. Eternity politicians spread the conviction that government cannot aid society as a whole but can only guard against threats. Progress gives way to doom. . . . eternity politicians manufacture crisis and manipulate the resultant emotion.
>
> *(Snyder, 2018, p. 13)*

Is the manufactured crisis at the southern border of the United States and the call for the construction of a border wall not exactly what Snyder is speaking about? The politics of inevitability and the resulting notion of eternity must not be allowed to creep into democratic society. If it gains a foothold, it will make changing the world-system for the better difficult, if not impossible. The politics of inevitability is cynical and will deter the changes needed in the present system, and the predicted doom will become reality.

John Shaar (1989), on the other hand, provides a more optimistic view when he outlines the open-endedness of the planning process. That view stresses that planning first begins with imagining the desired future and then in designing steps to achieve it. The future is not teleological, that is, inevitable, but one that is open-ended and created; it is just a matter of who is involved in its creation and how bold their imagination can be.

The social system, which includes the present environment, is in a chaotic state. I include the environment in the social system because of the manipulation and destruction of it by humanity. Systems by nature are naturally self-organizing and seek order from chaos. It is becoming evident that the present world-system has broken down. This chaos became increasingly evident during the COVID-19 virus. In fact, the virus may be the catalyst that clarifies just how inadequate and dysfunctional the present world-system has become. That is not to suggest that it was only the COVID-19 virus that identified the chaos in the system. The problems identified in Part 1 of this book have been demonstrating this chronic condition for some time. And, while some consideration has been given to these issues, there has not been sufficient attention devoted to creating political will and action to change the orientation of the world-system to address them effectively. The COVID-19 virus may be the large slap across the face that wakes humanity sufficiently to cause them to act, restoring order from the chaos. It is argued here that humanity is poised to search for a new world-system that will be better equipped to deal with the issues presented in Part 1. Part 2 has attempted to provide approaches and processes to tackle the change to the world-system, thereby moving civilization to the next stage, assuring continuance of the species. It is evident that present society is in a state of despair about its social and environmental situation. But as Eisenstein notes:

> [F]rom despair comes surrender, and from surrender comes an opening to new beliefs, a new conception of self and the world. From this might come a new way of relating to the world, a new mode of technology no longer dedicated to objectification, control, and eventual transcendence of nature.
>
> The collapse we are facing is of more than 'our civilization' but of *civilization* itself, civilization as we know it. It is a collapse of a whole way of relating to the world, a whole way of being, a whole definition of self.
>
> *(Eisenstein, 2007, p. 50)*

Creating a new world-system is a daunting contemplation. It can only happen when the number of variables in the present system are changed in concert so that the entire system transitions from chaos to sustainability. There are fundamental conditions that need to be operative for this transition to proceed and have any chance of being successful. First is when there is trust in the system's guardians who demonstrate that it clearly recognizes the issues important to the population and is dedicated to addressing those same issues. Additionally, those leading the change project, whether politicians or legitimate civic leaders, must demonstrate they are dedicated to providing an atmosphere that is conducive to conducting sincere and open dialogue. Finally, the population must utilize the cognitive tools that make the species unique, that is, an awareness of self and the ability to examine on mass the crises that befall us and hopefully come to the conclusion that the economic growth model that is on steroids at the moment, and has been for some time, is antithetical to the continuance of the human species on this planet.

This book was completed three months into the advent of the COVID-19 pandemic. There are events that come along from time to time that can act as a catalyst for big changes to be made to a system. This may be one of them, or it may just be another blip in history. Regardless of how it turns out, the COVID-19 pandemic has provided society and its leadership with a window of opportunity to examine what kind of a system civilization has created and how it might be made better. Since the pandemic has put the predatory corporate capitalist system on temporary hold, it is possible to step back and examine how we have been conducting human affairs over the recent past. Any knowing observer would agree that the 'growth on steroids' approach we have been taking to human affairs since the end of World War II is no longer sustainable if it ever was. As rational as this sounds, noise is already rising in the town square trying to distract us from that contemplation. Already we hear voices encouraging us to go back to our previous growth on steroids lifestyle, and in most cases prematurely. The elites and their surrogates are concerned that if we come to understand how the predatory capitalist system is destroying our social and environmental world, we may not want to return to what they describe as 'normal'. Their mantra is, 'let's get back to normal', but the exaggerated growth model of capitalism is not normal but an aberration of what and who we are or can be. Let's hope that this break in our frenzied chaotic world will give us pause to take stock and replace our worn out economic system with something that is more environmentally sustainable and socially just. Let's hope that the meaning of Homo sapiens (wise man) is justified in the end.

Bibliography

Bregman, R. 2017. *Utopia for Realists: How We Can Build the Ideal World [Kobo version]*. Kobo.com.

Eisenstein, C. 2007. *The Ascent of Humanity: Civilization and the Human Sense of Self [Kobo version]*. Kobo.com.

Elliot, L. 2018, January 22. Inequality Gap Widens as 42 People Hold Same Wealth as 3.7 bn Poorest. *The Guardian*.

Etzioni, A. 1997. *The New Golden Rule: Community and Morality in a Democratic Society*. New York: Basic Books.

Giddens, A. 1987. *Social Theory and Modern Sociology*. Cambridge, UK: Polity Press.
Giroux, H. A. 2018. *American Nightmare [Kobo version]*. Kobo.com.
Habermas, J. 1989. *The Theory of Communicative Action*. Oxford: Polity Press.
Hedges, C. 2018. *America: The Farewell Tour [Kobo version]*. Kobo.com.
McLuhan, M. 1967. *The Medium is the Message: An Inventory of Effects*. London, UK: Penguin.
McMurtry, J. 1999. *The Cancer Stage of Capitalism*. London: Pluto Press.
Pentland, A. 2014. The Rational Individual [Kobo version]. In *The Idea Must Die: Scientific Theories That Are Blocking Progress,* edited by J. Brokman, 597–601. Kobo.com.
Pimentel, D. A., Aymar, I. M., and Lawson, M. (2018, January). *Reword Work Not Wealth*. Oxford: OXFAM.
Polanyi, K. 1944. *The Great Transformation: The Political and Economic Origins of Our Time*. Boston: Beacon Press.
Reid, D. G. 2017. *Social Policy and Planning for the 21st Century: In Search of the Next Great Social Transformation*. London, UK: Routledge.
Shaar, J. 1989. *Legitimacy in the Modern State (2nd ed.)*. New Brunswick, NJ: Transaction Publishers.
Smith, A. 1771. *An Inquiry into the Nature and Causes of the Wealth of Nations*. London, UK: W. Strahan and T. Cadell.
Snyder, T. 2018. *The Road to Unfreedom [Kobo version]*. Kobo.com.
Streeck, W. 2016. *How Will Capitalism End? [Kobo version]*. Kobo.com.
Swingewood, A. 1991. *A Short History of Sociological Thought, 2nd ed*. London: MacMillan.
Wallerstein, I. 2004. *World-Systems Analysis: An Introduction [Kobo version]*. Kobo.com.
Wallerstein, I., Collins, R., Mann, M., Derluguian, G., and Calhoun, C. (2013). *Does Capitalism Have a Future? [Kobo version]*. Kobo.com.

EPILOGUE

In this book, I have presented what I believe to be the most significant problems facing humanity today and methodologies, processes, strategies, and actions that could be employed to address those issues. These problems affect the human community negatively and could spell the end of human life on this planet if they continue to be ignored. Some might argue this assertion is hyperbolic and that humans have faced severe crisis in the past and prevailed in the end, even flourished. The proof of that argument is the fact that we are still here and for most of us existing at a high standard of life. Even so, more and more people today view, with considerable alarm, what is going on in the world. The environment, particularly global warming, and social injustice continues to deteriorate significantly, and there does not appear to be a concerted worldwide effort to address these concerns. Some, including myself, would argue that social and environmental progress has stalled, perhaps even in decline.

During the course of writing this book, the world has been subjected to the COVID-19 virus pandemic. In addition to the health issues this pandemic has caused, it has also had devastating consequences to the capitalist economy and disruption to market-based work and the income it provides to individuals. For the most part, governments have come to society's aid and provided basic resources to those in need but who knows how long this support can last. The COVID-19 virus has demonstrated just how ill-suited our present form of capitalism is in addressing the needs of humanity. Our financial system – predatory corporate capitalism – has had to be nationalized because it was not equipped to cope with this type of catastrophe. Trillions of dollars (euros in Europe and pound sterling in Britain) has been pumped into the system by governments to prop up the capitalist economy rather than the economy coming to the aid of the population in this time of crisis. The United States, European Union, and other governments around the world have had to rescue their economies, because the capitalist system has not been able to cope with the basic needs of people during this upheaval but is now only designed to increase the wealth for those with wealth. World governments have been forced to intervene in the financial system to literally save lives. This is the second time since 2000 that governments have had to bail out the capitalist system to keep it from collapsing. Hopefully, the world has learned just how important government is to the functioning of society, dispelling the myth that government is 'the problem', and a drag on social functioning and the private marketplace is the best, perhaps only mechanism for achieving social justice. The pandemic has demonstrated the vulnerability of this myth and the need for world-system change. The present world-system has demonstrated itself to be no longer suited to solving many of humanity's pressing problems but is antithetical to continued social progress and environmental sustainability. It is now definitely time to heed Polanyi's warning and embed the economy in society rather than society in the economy.

I wrote this book during the Donald J. Trump US presidency and the British withdrawal from the European Union. Certainly, Trump's presidency profoundly changed the tone of domestic and geopolitics. That said, I believe that Trump is the outcome of forces that are deeper and more significant than the instigator of what has transpired since his inauguration. There is great anxiety in society today, and the oligarchs are taking advantage of the situation, stoking those fears in order to increase their hold on society. That said, there is resistance growing that is concerned with the deterioration in the environmental and social system. Additionally, the COVID-19 pandemic has produced cause for society to reflect on the present condition and question where society seems to be headed.

There is today, and has been for some time, a culture war going on in the United States and other parts of the western world. Some of the difference between people is the traditional right-left political divide, but there is a closed/open society debate, as well. There is a growing populist sentiment that leans to isolationism and nationalism and is xenophobic. As you might expect, there has been a reaction to this xenophobia as shown by the strengthening of the Black Lives Matter movement. We are witness to the rise of autocracy in many of what we generally think of as liberal democratic countries. This mass anxiety is not simply a result of the alienation produced by economic globalization that has negatively affected the economic middle class and poorer members of society, but a shift in basic social tenor. Much of this hard-right turn in social attitudes is stoked by the social instruments supported by those at the top of the economic ladder. On the other hand, or perhaps as a reaction to the ultra-right push, many citizens are shrugging off their complacency and beginning to assert themselves with the intention of redressing the imbalances that have occurred in the environment and social justice system.

Those at the top are concerned that society is becoming too liberal and they are making a play to take over government. The Obama administration, along with other democratic governments in the European Union, Canada, and elsewhere in the world, initiated the beginnings of broadening the globalization project that focused more on human rights and less on economic exploitation. It seems like a long time ago, but it is worth remembering that it was Obama who gave strong leadership to the Paris Climate Accord initiative. These progressive initiatives engendered fear in the wealthy that a steady erosion of their unfettered freedoms through increasing regulations and introduction of stronger social programs was beginning. On their behalf, Trump has virtually erased many of the environmental regulations enacted by previous governments and is regressing to the 1950s environmental laws that allowed companies to spew all types of unhealthy toxins into the environment before their deleterious effects were known to science. Fortunately, other governments throughout the world have not followed suit, but progress on many environmental and social issues has slowed down. Most importantly, carbon pollution is still increasing, endangering the climate of the planet. The world is definitely warming up.

Deregulation is taking place in other spheres of social and political life that will take us back to, at least, the 1950s or maybe earlier. In order to maintain power, those in charge of the political system are making voting more difficult for minority groups that often don't support their ideology. Women's rights are now in jeopardy. The supreme court in the United States is now made up of justices that are not the friends of such issues as a women's right to choose whether to carry a foetus to term or not and other such ideological issues. What used to be the political centre has moved dramatically to the right. Certainly, there are voices that are attempting to return the left to the left, but whether they will be successful or not remains to be seen.

There are now many international organizations funded by a select group of the very wealthy whose goal is to increase their position in the world at the expense of the rest of society. The wealthy are creating or funding many right-wing non-governmental organizations to saturate society with their propaganda and to do their bidding. They go about this by spewing propaganda about the existence and nefarious intents of a Deep State and other untruths masquerading as fact. Lying, otherwise known as alternative facts, has become a major communications strategy of this group. They pour billions of dollars into an effort to control government and society.

Most books like the one provided here present abstract notions, such as social justice, equality, democracy, and so on. It is often difficult for people in society to relate to such abstractions even though no one can argue against their importance. What I have done here is to provide concrete issues that affect all of us dramatically and directly, and those which people experience personally. We see the results of climate change in the wildfires on the Pacific coast and the drastic flooding in other parts of the country. These same events are occurring over the entire world. The increase and severity of the number of hurricanes we encounter annually cannot be dismissed as status quo events. The globe is warming and making the repercussions apparent to everyone even though some want to deny it. Likewise, there is a recognizable deterioration in the social justice system that rivals the severity of environmental concerns. Further, there are many groups acting on these issues individually, so there is a readiness to see the interconnection and conjunction of these single issues as they relate to world-system change. As stated before, no single issue will be resolved without the resolution of all of them together, which will necessarily lead to world-system change.

The opposition to system change, and the preservation of the status quo, continues today. There are many private think tanks around the world whose sole purpose is to perpetuate the present system regardless of its danger to humans and the environment. These international organizations whose membership consists of billionaire libertarians wish to maintain a world order that is top down and controlled by those with wealth. They are not warm to the idea of an egalitarian society but much more attuned to the Spencerian notion of survival of the fittest. They help organize resistance in all countries that have adopted a welfare state solution to human problems and fund country-specific organizations that are opposed to social justice reform. Perhaps the experience of the COVID-19 virus will impact that train of thought.

Most opposition movements that are attempting to counter the propaganda thrust, described earlier, use science and fact to counter the ultra-libertarian argument. The use of facts in forming an argument that describes the world is undeniably important. To counter the scientific facts, the alt-right has vilified the purveyors of these facts, the media and others, as perpetrators of false news. They are described as the enemy of the people. They have coined the term 'alternative facts' to sanitize false statements and out-and-out lies.

The counterargument to right-wing populism needs to create and adopt an emotional story of how the world ought to be using a facts-based approach. People need to feel an emotional attachment to any narrative that purports to guide their future development. The alt-right have understood that need and have provided such a narrative that highlights grievance, fear, and anxiety. Historically, we have witnessed the creation of emotional attachments to social change that have been very successful, if not always desirable. Hitler's Nazi Germany is one great example. The German population was emotionally moved by Hitler's rhetoric and narrative for the future for a struggling Germany after WW1. Certainly, Churchill also created a compelling narrative that described Britain as the underdog in World War II that would not be defeated. Churchill provided the British people with a narrative of hope that appealed to their emotional concept of themselves. The democratic society has neglected to create an emotional story to outline its many positive features, even though it has employed the use of a fact-based argument. We drastically need a narrative that has great emotional appeal. Surely the survival of the planet and all life on it can provide that kind of emotional appeal.

INDEX

3D printing 37, 42–43

action 18, 21, 30, 45, 67–68, 76, 88, 129, 165, 194, 198, 204–205, 208, 210; nefarious 12, 20–22, 42, 94, 96, 111, 113; process 7, 128; rational 23, 168–172, 175, 177, 179–181, 185, 187, 190; remedial 32–34, 53, 67, 71, 83, 104, 132, 135, 137, 144–149, 158, 161, 167
agency 181, 189, 206
Alston, P. 58
anger 6, 9, 12, 77, 105, 110, 124–126, 138–139, 176, 200, 210
Antarctic 18, 25, 27
anywheres people 77–78; see also somewheres people
Appadurai, A. 74, 82–83
Arctic 15, 17–18, 20, 24–28
Artificial General Intelligence 37, 45–46, 49–53
artificial intelligence 36–39, 42–46, 48–49, 53, 73, 79–80, 143, 196, 198, 200
authoritarian 126, 198, 208–209; governments 5, 10, 77, 103, 112; leaders 66, 80, 116, 130, 158; regime 7, 117

Barber, B. 98
basic needs 8, 138–140, 158, 164, 169, 216
Bertalanffy, K. L. 146
Black Lives Matter (BLM) 70, 127, 183, 185
Boulding, K. 146

Bremmer, E. 66, 78, 115
Brexit 12, 68, 75, 78–80, 94, 121–126, 130, 137, 172, 200, 203, 206
Bryne, D. 148–150

Canadian Index of Wellbeing 65
capacity building 176, 185
capitalism 4, 7–8, 19, 31, 50, 59, 70, 74, 85; capitalist system 10–11, 20, 51, 54, 63, 78, 137, 151, 154–155, 160–164, 196, 199, 202; capitalist world-system 60, 76, 89–90, 134, 143, 162–164, 176–177, 198; decline 3, 5, 156, 188, 193, 197–198; see also preditory corporate capitalism
Carroll, S. 168
climate change 15, 21, 24, 31, 209; deniers 17, 20, 112, 114; effects 17, 23, 27, 36; evidence 19, 22, 150; political issue 195, 197, 206, 211, 217; reports 23–29
CO_2 20, 24–25, 29, 33, 109, 114, 191; reduction 17, 19, 23, 30–32, 190, 204, 207
coastal cities 16, 25–26
cognition 45–46, 136, 186
common ground 78
community 11, 25, 47, 103, 201; anti 112; business 30–31, 74, 76; development 64, 83, 185–187; education 172, 178; intelligence 41, 118; organization 8, 54, 133–134, 140, 174; scientific 15–17, 26–27, 31, 44, 51; world 80, 85, 88, 90, 92

220 Index

cooperative structure 4, 6, 20, 31, 55, 81, 85, 110–111, 138, 197, 206
corporate capitalism 3, 6, 55, 62, 73, 76, 90, 135, 147, 155, 173, 194, 196–198, 202, 206–208, 213; *see also* predatory corporate capitalism
Covid-19 4, 20, 67, 94, 101, 114, 116, 125, 135, 139, 187, 197–199, 204, 207, 212, 215
critical theory 157–158, 164, 175; communicative action 175, 179, 205, 208
crop failure 29
culture 6, 9, 19, 38, 53, 73, 80, 84–85, 96, 99, 104, 124, 140, 153–154, 177, 190; clashes 13, 74, 91, 93–95, 101–105, 130, 207; wars 86, 90, 97–98, 216
cyberspace 40–41, 74

decision-making 16, 51, 125–126, 164; authorities 120; bottom up 146; collective 88, 136, 210; evidence-based 124, 133, 170
deglobalization 77, 87, 89, 197, 201
democracy 7, 10, 41, 56, 124, 127, 133, 136, 163, 167, 180, 183, 193, 206; liberal 65, 70, 75–77, 109, 111, 118, 120, 161, 208; social 66, 196, 198, 200, 202
demographic change 57, 66, 88, 101–103, 129
developing world 56, 59, 70, 76, 86–87, 92, 113, 199
dialogue 123, 128, 141, 146, 156, 174, 181, 184, 194; framework 146, 179, 213; inclusive 51, 78, 136, 169, 170–174, 179–180, 196, 205
driverless vehicles 36, 42–44
drones 42
Dyer-Witheford, N. 35, 46, 48
Dynamic Integrated and Regional Integrated Climate Model 29

economic classes 10, 56–58, 62, 89, 110, 115, 119, 122, 124, 127, 137, 155, 194, 196, 207; economic growth 20, 31, 68–70, 75, 87, 99, 109, 204, 213; economic system 4–12, 24, 33–34, 36, 38, 43, 50, 63, 65–66, 74, 76–77, 81, 83, 88, 100, 115, 134, 143, 161–163, 170, 173–177, 181, 183, 188, 193, 199
education 51–52, 64, 79, 173, 178; education and communications program 21–22, 104, 170–172, 180, 184, 187, 209; experiential 83, 136, 172, 174, 180
Eisenstein, I. 6, 15–16, 90, 212

emancipation of humans 38, 158, 164, 174, 205–206
emotional intelligence 6, 16, 77, 124–125, 144–145, 164, 188, 200
Enlightenment 133, 142–143, 158
environment 4–5, 7, 10, 19, 30, 78, 83, 88, 132, 184, 190, 197, 204; change 9, 16, 43, 68, 81, 133, 164, 174, 182, 188, 208; damage 3, 7–8, 11, 20, 24, 28–29, 34, 37, 90, 96, 114, 123, 125, 140, 153, 169, 179, 202, 213; projections 131, 134, 143, 154, 167–168, 211; social 6, 84–85, 98, 156, 177–178, 194, 212
ethics 48, 51–52
Etzioni, A. 198
evolution 13, 16, 31, 38, 53, 126, 134, 136, 153, 169, 177, 188
exploitation 35, 62, 86

false consciousness 128, 182; market glorification 170, 198
fear and anger 12, 77, 105, 124–125, 140, 211
feedback 136–137, 151, 170, 180–181, 196
financial corporate complex 195
Flannery, T. 18–19
Frank, R. 49, 66
Frankl, V. 50
Freeland, C. 173–174
Freire, P. 174
Friedmann, J. 175–176, 185
Fukuyama, F. 10, 56–57, 131

Galbraith, J.K. 143, 154
genes 38–39, 44, 48
genetic research 44
Geuss, R. 158, 175
Giddens, A. 143, 157–158, 174–176, 181, 184, 205–206
Giroux, H. A. 5, 8, 77, 196, 201, 206, 208
global capitalism 74
Global Compact 93
global framework 27, 197, 211
globalization 4, 8, 11, 13, 35, 66, 73–82, 84, 86, 88, 110, 122, 124–125, 201, 216; bottom up 82–85, 88–89
global warming 9–10, 13, 15–17, 19–47, 49, 54, 92, 109, 114, 119, 125, 191, 202, 215; mitigation 81, 150, 153, 155, 182, 188, 194
Goodhart, D. 77–78, 88
grassroots institutions 83
Green New Deal (GND) 71, 127, 155, 173, 180, 197–202
grievance society 111, 181, 201–202, 206, 208

Gross Domestic Product (GDP) 10, 57, 64, 66, 76, 81, 115, 176
guaranteed income 60, 71, 139, 199, 200, 204

Haass, R. 80–81, 89
Habermas, A. 88, 158, 170, 176, 179, 205, 208
Hanson, J. 15, 27–28
Harvey, D. 8, 132
healthcare 60, 63, 197, 204, 206
Hedges, C. 57, 59, 84, 87, 196
Hobbes, T. 143
Homo sapiens 39, 48; Homo techno 48; wise man 186, 213
human rights 58, 75, 77, 83, 85, 121, 125, 197, 202

ice caps 18, 20, 25, 28
immigration 57, 66–67, 85, 92, 99, 109, 121; anti-immigration 12, 86, 93–94, 100–104, 115; resistance 122, 130
income 11, 50, 57, 61, 63–65, 69, 79, 139, 167, 199, 202; unequal distribution 3, 58–59; world statistics 10, 65, 68, 88
increasing returns 151
individualism 175, 198
inequality 5, 7, 54, 57–59, 64–71, 82, 86, 124, 127, 176–177, 188, 194, 196, 207; limited 177
institutional determinism 181
integration 13, 75, 82, 88, 103–104, 120, 126, 186, 197, 207
Intergovernmental Panel on Climate Change (IPCC) 13, 22–27, 31, 33
isolationism 13, 80, 85, 87–88, 90, 110–111, 116, 119, 122–123, 182, 193, 197, 202, 208, 210

Jeffko, W. 112

Keynes, J. M. 162–163
Krugman, P. 163
Kuhn, T. 152, 154, 192; anomaly build up 155–156; paradigm shift 164, 176
Kuttner, R. 167

leaders 30, 33, 54, 60, 68, 138, 172, 180–181; corporate 85; political 51, 78, 104, 113, 116–117, 201
leadership 185, 200, 208, 210, 213; abdication 22, 119; Chinese 116; world 20–21, 80, 105, 110, 135, 156, 158, 183–184, 187
leisure 50, 52, 65, 77

life-world 87–89, 132, 135, 158–159, 161, 164, 170, 177, 197, 202, 204–206; *see also* system-world
livable wages 58, 62–65, 67, 76, 79, 86, 129, 198
Locke, J. 143
Luce, E. 65

Mair, H. 52
manufacturing 7, 11, 30, 36–37, 42–43, 49, 63, 49, 63, 66–63, 75, 79, 81, 86, 109, 112–113, 138, 143, 161, 199, 201
marginalization 77, 137, 184, 206; marginalized 6, 12, 56, 61, 64, 77, 84, 88, 109–111, 118–119, 121, 123, 137, 143, 169, 200, 202, 210
Marx, K. 7, 138, 143, 162, 196, 198
Maslow, A. 139–140, 158, 160–161, 164, 199, 202
Mayer. J. 5
Mazzucato, M. 55
McCann, K. 186
McClelland, J. S. 112
McLuhan, M. 197
Media Celebrity Complex 205
Metzl, J. 44
migration 9, 84, 92, 105, 121, 190, 207; climate related 92, 184; escaping war 101; Muslim 91–95, 100; South America 91, 99
Mueller, R. 116–117

NAFTA (USMCA) 12, 75, 122, 197, 201
nationalism 12, 50, 57, 62, 66, 77, 80–83, 87–90, 92, 100–102, 104, 110–111, 116, 119, 123–126, 160, 182, 190, 193, 199, 200–203, 208, 210–211
natural selection 44, 54, 136
neo-liberal democratic world 111
new world-system 4, 7, 13, 78, 90–91, 119, 134, 136, 139, 156, 167, 175, 177, 179–180, 182–183, 186–187, 189, 193, 196–197, 199, 201–202, 204, 208, 210, 212
Nordhause, W. 29–30

Occupy Movement 70, 127, 185
oligarchs 5, 63, 158, 185, 194, 196, 198, 202, 205, 209
organizational development 176
outreach strategy 174
overarching theme 71

paradigm 6, 150, 152–156, 164, 170–171, 173, 200, 202, 209, 211; *see also* Kuhn, T.

Index

Paris Climate Accord 15, 19, 21, 30, 81, 112, 114, 120, 122, 188, 209–210
participation 33, 52, 63, 79–80, 85, 124, 175–176, 179, 186, 208–209
Pentland, A. 145
permafrost 24; melting 18, 26–28
Piketty, T. 56, 102–103
planning 122, 145, 149, 169, 181, 188, 190, 204, 209; bottom up 85, 175; definition 136, 168, 212; framework 179, 182, 186; human activity 45; process and stages 170, 173, 175–179; taskforce 180, 182, 188–189; transactive 168–170, 174–175, 178, 210; turning ought into is 131–132, 149, 169, 218
Plutocracy 11, 33, 58, 126–127, 164
Polanyi, K. 76, 162–164, 202
policy 8, 31, 48, 60, 79, 121, 139, 167, 169–170, 173, 188, 196–197, 202, 205; anti-immigration 100, 104, 115, 208; anti-social 187, 194, 202; assimilation or multi-culturalism 103–104; deconstruction 114; economic 7, 62, 81, 118, 127, 162, 199, 202; environmental 21, 28, 32, 34, 81, 85, 114, 188; evaluation 170, 177, 180; imigration 102, 104–105, 115; implementation 171, 182, 184, 208; isolatonist/nationalist 4, 88, 100, 122, 126; public 5, 117–118, 121, 128, 133, 135, 137; rational 132, 169; social 13, 57, 60–61, 70–71, 109, 136
political 5, 7, 12–13, 21, 29, 31–35, 41, 53, 78, 80, 83, 89, 93, 97–99, 101–104, 110–121, 123, 127–130, 132, 135, 139, 168, 172–173, 206, 210–211; gridlock 169, 177, 183, 185, 194, 197, 205–206, 208
pollution 20–21, 29, 31–32, 81, 201, 207
populism 49–50, 57, 62, 66, 77, 87–89, 110–111, 120, 123–125, 138, 160, 193, 201, 207–211
poverty 32, 44, 53, 58–59, 61–65, 68, 76, 84, 91–92, 99, 100–101, 105, 125, 154, 188, 190–191, 207, 211; absolute 56, 69–70; eradication 60, 71, 136; reduction 57; relative 56, 70, 155
power 3–4, 11, 20, 32–34, 61, 69–70, 82–84, 86–88, 89, 94, 97, 112, 115–116, 120, 130, 134, 137, 145, 164, 167, 173, 185–186, 206, 208, 210–211; politics 4, 6, 11, 20, 77–80, 90, 109, 116, 119, 169, 187, 206, 208–221; religion 97; sharing 84, 128; struggle 5, 53, 110, 145, 200

predatory corporate capitalism 6, 60, 64, 66, 92, 118, 167, 183, 194, 207, 213
probability 48, 150, 152, 165, 182
public education 27, 33, 51, 170

racism 88, 102, 112, 114, 193; *see also* xenophobia
rational thought 82, 200; process 124–126, 168, 210
Redner, C. 74
reductionist analysis 146; research methods 144, 154; *see also* scientific method
regulations 81, 180, 188; environmental 19, 31, 114; technology 48
Reid, D. G. 8, 50, 136, 186, 196, 199
Reid, D. G. and Van Dreunen, E. 185
religious fundamentalism 84, 87
Rich, N. 15, 21, 27
robotics 36, 38, 41, 50, 67–68, 86, 137, 143, 198
Rothman, J. 187
Rousseau, J. J. 143

Sachs, J. 65
Saleebey, D. 181
Sanger, D. E. 44
Schumpeter, J. A. 4, 138
scientific method 132, 146, 154–155, 168; *see also* reductionist analysis
sea levels 15, 17–18, 23, 25–27
Selye's General Adaptation Syndrome (GAS) 152–154, 164, 188
Shaar, J. 178, 212
Skidelsky, R. & E. 69
Skyttner, L. 146, 149
Smith, A. 70, 143, 162, 194
Snyder, T. 6–7, 9, 211
social 5, 6, 28, 46, 56, 62–64, 68–71, 92, 135, 148, 157, 185; alienation 114, 118–119, 123, 186, 193; anomie 36, 138, 184, 188; architecture 14, 35, 129, 133, 136, 138, 160, 169, 171, 193, 197, 199–200, 205; breakdown 37, 169, 179, 200, 202; change 9, 36, 62, 113, 129, 131, 141, 162–164, 170, 173, 182, 184; cohesion 37, 49, 113, 144, 169; contract 11, 13, 30, 83, 127, 130, 170–171, 178, 194, 200, 210; democracy 66, 87, 196, 198–200, 202; justice 4, 9, 13, 83, 87–88, 109, 125, 131, 137, 154, 169, 170, 174, 177, 181–182, 187–188, 197, 202, 206–207; learning 136–137, 174, 176, 189, 196; media 13, 40–41, 53, 128, 182; movements 128, 140, 186, 205; solidarity 5, 63, 65, 75, 88, 98, 103, 127, 137–139, 172, 177, 204;

Index **223**

system/structure/organization 3–7, 11, 16, 21, 34, 36, 48, 50, 54, 61, 66, 73, 80, 85, 95, 126, 132, 142, 145, 153–154, 156–158, 163–165, 168–171, 173, 177, 181–182, 185, 196, 199, 207, 211–212; theory 7, 142, 179; three stage process 173; transition 8–10, 38–39, 110, 143, 171, 182, 210; unrest 66, 200–201, 204
somewheres people 77–78; *see also* anywheres people
Srnicek, N. 39
Stebbins, R. 52
Streeck, W. 3, 196
sustainability 7, 83, 132, 150, 170, 197, 213; environmental 154, 177
Swingewood, A. 157, 205
symbolic communication 123
system-world 87–88, 90, 132, 141, 159, 162, 195, 204–206; rebalance with life-world 135, 137, 164, 170, 182, 197, 202–203, 208; *see also* life-world
systems theory 81–82, 136–137, 152, 164

Tauber, A. 45
tax 60, 65; carbon 21, 30, 32, 190, 204; marginal rate 59, 71, 202; robots 67, 200; tax cut 4, 118, 198
technology 6–9, 11, 13, 38, 44–45, 47, 55, 57, 63–64, 67, 77, 80, 88, 110, 183, 207, 211–212; digital 37, 53, 77, 79, 131, 142, 161; green 30–33; mechanical 10; new 36, 46, 68, 104, 114, 132, 138, 143, 156, 199; rapid 4, 177; security 51, 54, 81, 196; symbiotic 49
Tegmark, M. 44, 47, 50
temperature 8, 18, 24, 31; air 23; earth 20, 32
theory 7, 13, 54, 70; centre/periphery 99; chaos and complexity 47, 150–153, 164; *conjuncture* 134–135, 152, 165, 207; *contingent* 134, 152, 191; critical 158, 162, 164, 175, 205; conomic 162–163; evolution 16, 31, 37–38, 53, 73, 136, 153, 168, 177, 188; Giddens Theory of Structuration 157, 164, 175, 182; heuristics 144; industrialism/administrative 143; Kuhn's paradigm shift 152, 154; Maslow's theory of human motivation 158, 160–161, 164; *random accident* 134, 152; structuralism 156–157; systems 81–82, 136–137, 146–147, 152, 164; theory of mind 123–124; trickle down 11, 83–84, 155, 164
Trans Pacific Partnership 112, 120

unequal wealth distribution 3, 9, 11, 59, 68–69, 125, 174, 191, 205

value 11, 22, 31, 50–51, 69, 75, 77, 89–90, 98, 110, 117, 131, 133, 144, 168–169, 171, 173, 177, 179, 183, 189, 209–210; instrumental/esthetic/moral 145–146
value criteria 157, 177, 182, 209
Vance, J. D. 61–62

Waldrop, M. M. 136, 149
Wallerstein, I. 3–4, 8, 19–20, 110, 144–145, 192–193
weather patterns 17–19, 22, 24–29, 92
wellbeing 50, 62, 64, 69–71, 81, 161, 174, 206
Wharf, B. and McKenzie, B. 169, 177, 181
Wilkinson, R. & Pikett, K. 56, 69–70
Wilson, E. O. 156
work 11, 36–40, 42, 44–46, 49–51, 54, 63–64, 67, 75–78, 104, 139; meaningful 49, 161; workplace 11, 37, 44, 49, 51, 57, 64, 67, 138, 160
World Health Organization (WHO) 12, 47, 179

xenophobia 12, 57, 66, 92, 100, 104, 113, 121, 125; *see also* racism

Taylor & Francis eBooks

www.taylorfrancis.com

A single destination for eBooks from Taylor & Francis with increased functionality and an improved user experience to meet the needs of our customers.

90,000+ eBooks of award-winning academic content in Humanities, Social Science, Science, Technology, Engineering, and Medical written by a global network of editors and authors.

TAYLOR & FRANCIS EBOOKS OFFERS:

- A streamlined experience for our library customers
- A single point of discovery for all of our eBook content
- Improved search and discovery of content at both book and chapter level

REQUEST A FREE TRIAL
support@taylorfrancis.com